The IEE

Protection Against Overcurrent

GUIDANCE NOTE 6

IEE Wiring Regulations

6

BS 7671 : 2001 Requirements for Electrical Installations

Including Amd No 1 : 2002

Published by: The Institution of Electrical Engineers, Savoy Place,
LONDON, United Kingdom. WC2R 0BL

Printed January 1993
Reprinted August 1993, with minor amendments
2nd edition July 1996, incorporating BS 7671 : 1992 inc Amd No 1
3rd edition Aug 1999, incorporating BS 7671 : 1992 inc Amd No 2
4th edition Sept 2003, incorporating BS 7671 : 2001 inc Amd No 1

Copies may be obtained from:

The IEE
PO Box 96, STEVENAGE,
United Kingdom. SG1 2SD

Tel: +44 (0)1438 767 328
Fax: +44 (0)1438 742 792
Email: sales@iee.org
http://www.iee.org/publish/books/WireAssoc/

ISBN 0 85296 994 5, 2003

Contents

Co-operating Organisations

The Institution of Electrical Engineers acknowledges the contribution made by the following organisations in the preparation of this Guidance Note.

British Cables Association
    J M R Hagger BTech(Hons) AMIMMM

British Electrotechnical & Allied Manufacturers Association Ltd
    P D Galbraith IEng MIIE MIMgt
    R Lewington MIEE

British Standards Institution
    M Danvers

CIBSE
    Eur Ing G Stokes BSc(Hons) CEng FIEE FCIBSE

City & Guilds of London Institute
    H R Lovegrove IEng FIIE

Electrical Contractors' Association
    D Locke IEng MIIE ACIBSE

Electricity Association
    D J Start BSc CEng MIEE

Engineering Equipment and Materials Users Association
    R J Richman

EIEMA
    Eur Ing M H Mullins BA CEng FIEE FIIE

ERA Technology Ltd
    M W Coates BEng

Health & Safety Executive
    Eur Ing J A McLean BSc(Hons) CEng FIEE FIOSH

Institution of Electrical Engineers
    G D Cronshaw IEng MIIE (Editor)
    P R L Cook CEng FIEE MCIBSE
    P E Donnachie BSc CEng FIEE

National Inspection Council for Electrical Installation Contracting

Office of the Deputy Prime Minister
    E N King BSc CEng FCIBSE

Royal Institute of British Architects
    J Reed ARIBA

SELECT / ECA of Scotland
    J Davidson CEng FIEE
    D Millar

Acknowledgements

References to British Standards are made with the kind permission of BSI. Complete copies can be obtained by post from:

BSI Customer Services
389 Chiswick High Road
London   W4 4AL

Tel:	General Switchboard:	020 8996 9000
	For ordering:	020 8996 9001
	For information or advice:	020 8996 7111
	For membership:	020 8996 7002
Fax:	For orders:	020 8996 7001
	For information or advice:	020 8996 7048

BSI operates an export advisory service - Technical Help to Exporters - which can advise on the requirements of foreign laws and standards. The BSI also maintains stocks of international and foreign standards, with many English translations.
Up-to-date information on BSI standards can be obtained from the BSI website http://www.bsi-global.com/

Copies of Health and Safety Executive documents and approved codes of practice (ACOP) can be obtained from:
HSE Books, P O Box 1999
Sudbury, Suffolk   CO10 6FS
Tel: 01787 881165   Fax: 01787 313995

Health and Safety Executive telephone enquiries can be made to:
HSE Info Line on : 08701 545500

Fax and postal enquiries can be made to:
HSE Information Centre
Broad Lane, Sheffield   S3 7HQ
Tel: 01142 892345   Fax: 01142 892333.

Preface

This Guidance Note is part of a series issued by the Wiring Regulations Policy Committee of the Institution of Electrical Engineers to enlarge upon and simplify some of the requirements of BS 7671 : 2001 inc Amd No 1, Requirements for Electrical Installations (IEE Wiring Regulations Sixteenth Edition). Significant changes made in this 4th edition of the Guidance Note are sidelined.

Note this Guidance Note does not ensure compliance with BS 7671. It is a guide to some of the requirements of BS 7671 but users of these Guidance Notes should always consult BS 7671 to satisfy themselves of compliance.

The scope generally follows that of the Regulations and the principal Section numbers are shown on the left. The relevant Regulations and Appendices are noted in the right-hand margin. Some Guidance Notes also contain material not included in BS 7671 but which was included in earlier editions of the Wiring Regulations. All of the Guidance Notes contain references to other relevant sources of information.

Electrical installations in the United Kingdom which comply with BS 7671 are likely to satisfy the relevant aspects of Statutory Regulations such as the Electricity at Work Regulations 1989, but this cannot be guaranteed. It is stressed that it is essential to establish which Statutory and other Regulations apply and to install accordingly. For example, an installation in premises subject to licensing may have requirements different from, or additional to, BS 7671, and the requirements will take precedence.

Users of this Guidance Note should assure themselves that they have complied with any legislation that post-dates the publication.

Introduction

This Guidance Note is concerned with Part 4 of BS 7671— Protection for Safety.

Neither BS 7671 nor the Guidance Notes are design guides. It is essential to prepare a full specification prior to commencement or alteration of an electrical installation.

The design and specification should set out the requirements and provide sufficient information to enable competent persons to carry out the installation and to commission it. The specification should include a description of how the system is to operate and all the design and operation parameters. It must provide for all the commissioning procedures that will be required and for the provision of adequate information to the user. This should be 514-09 by means of an operation and maintenance manual or schedule, and 'as fitted' drawings if necessary,

It must be noted that it is a matter of contract as to which person or organisation is responsible for the production of the parts of the design, specification construction and verification of the installation and any operational information.

The persons or organisations who may be concerned in the preparation of the specification include:

The Designer
The Planning Supervisor
The Installer
The Supplier of Electricity
The Installation Owner (Client) and/or User
The Architect
The Fire Prevention Officer
All Regulatory Authorities
Any Licensing Authority
The Health and Safety Executive

In producing the design and specification advice should be sought from the installation owner and/or user as to the intended use. Often, as in a speculative building, the intended use is unknown. In such cases the specification and/or the operational manual must set out the basis of use for which the installation is suitable.

131-01-01

Precise details of each item of equipment should be obtained from the manufacturer and/or supplier and compliance with appropriate standards confirmed.

511

The operational manual must include a description of how the system as installed is to operate and all commissioning records. The manual should also include manufacturers' technical data for all items of switchgear, luminaires, accessories, etc and any special instructions that may be needed. The Health and Safety at Work etc Act 1974 Section 6 and the Construction (Design and Management) Regulations 1994 are concerned with the provision of information. Guidance on the preparation of technical manuals is given in BS 4884 (Technical manuals, specification for presentation of essential information, guide to content and presentation) and BS 4940 (Recommendations for the presentation of technical information about products and services in the construction industry). The size and complexity of the installation will dictate the nature and extent of the manual.

Section 1 — The Regulations Concerning Protection Against Overcurrent

1.1   Scope

Fundamental principles

1.1.1
130

The fundamental principles include two regulations concerning overcurrent.

130-04
130-05

The first of these requires that, so far as is reasonably practicable, persons or livestock shall be protected against injury and property shall be protected against damage due to excessive temperatures or electromechanical stresses caused by any overcurrents likely to arise in live conductors.

The second regulation requires conductors and any other parts likely to carry a fault current to be capable of doing so without attaining an excessive temperature.

1.1.2

Protective measures

473
530
533

Measures to provide protection against overcurrents are contained in Chapter 43, Section 473, Sections 530 and 533.

Chapter 43 gives the basic requirements for both overload and fault current protection, whilst Section 473 has further requirements concerning the application of the protective measures and the location of the protective devices.

Sections 530 and 533 contain regulations for the selection and installation of the protective devices.

1.1.3   *Load equipment and flexible cords*

It must be remembered that the Regulations deal primarily with the fixed equipment of an installation. Protection of conductors in accordance with Chapter 43 does not necessarily protect the equipment connected to an installation nor the flexible cables and cords connecting such equipment to the fixed part of an installation. However, some notes are provided in Section 8 of this Guidance Note on the overcurrent withstand of flexible cords. This subject is of importance where BS industrial-type plugs and sockets are to be used. It may be necessary to consider the feasibility of a limitation on the current rating of such circuits, to introduce local protection or to require minimum conductor sizes for the flexible cords.

1.2   Nature of overcurrent and protection   *General*

Every circuit shall be protected by one or more devices which automatically interrupt the supply in the event of overcurrent.

1.2.1

The object of the Regulations is to ensure that any overcurrent does not persist long enough to cause damage to equipment or risk of injury to persons or livestock.   130-04
130-05
131-08

An overcurrent is any current which exceeds the current-carrying capacity of the circuit/ conductors.   Part 2

This is determined by the nominal current or setting of the overcurrent protective device for that circuit, and the current-carrying capacity of its conductors by Regulation 433-02-01. The conductors must also be protected against fault currents.   433-02-01

1.2.2   *Overload and fault current*

433   The term overcurrent includes both overload current
434   and fault current, see Sections 433 and 434.
Fault current may be short-circuit current or
earth fault current.

The difference, in principle, between overload current and fault current lies in the reason for their occurrence, not in their magnitudes. An overload is imposed on a sound circuit by an abnormal situation at the load, whereas fault current is due to an insulation failure or bridging of insulation in the circuit or its terminations.

In practice, however, the difference between the two types of overcurrent tends to be one of current magnitude and duration, because this makes for a convenient classification of the protective devices required. The total time/current characteristic of a protective device or combination of devices should provide protection for any combination of current and time which could result in overheating or mechanical over-stressing. This total characteristic is exemplified in the single characteristic of a fuse or the combined characteristic of a circuit-breaker with its thermal and electromagnetic trip mechanisms.

It is generally sufficient to look at the two extremes of this characteristic; the high current short duration part which, because it usually protects from faults in the circuit itself, is referred to as fault protection and the long duration lower current part, usually protecting against excessive load, which is referred to as overload protection.

Furthermore, experience shows that these two extremes cover practically all abnormal currents and where calculation of the values of current is the most practicable.

It is accepted, however, that actions such as direct-on-line motor starting, or capacitor switching, result in currents of a magnitude similar to some fault currents, while a high resistance earth fault or a fault in equipment could result in currents of the same magnitude as those arising from excessive loads.

1.2.3   *General protection characteristic*

433

The total characteristic of circuit protection complying with the Regulations is intended to be such that if the protection is adequate at the two extremes it will be satisfactory for any time/current combination in between.

For practical reasons it is sometimes not feasible to provide complete protection for very small overloads and attention is drawn to the comments in paragraph 2.2.5 on Regulation 433-01-01 with regard to load assessment.

433-01-01

1.3   **Statutory requirements 110-04**

The Electricity at Work Regulations 1989 require that efficient means, suitably located, shall be provided to protect against excess current in every part of a system, as may be necessary to prevent danger.

It is important that the designer and the installer consult the relevant documents listed in Appendix 2 of BS 7671, together with those Statutory Regulations and Memoranda which apply to the particular installation concerned.

Appx 2

Regulation 110-04-01 sets out the relationship of BS 7671 : 2001 with Statutory Regulations, including the Electricity Supply Regulations 1988 as amended which have now been superseded by the Electricity Safety, Quality and Continuity Regulations 2002. The Statutory Regulations are also listed in Appendix 2 of BS 7671 : 2001.

There may also be local regulations which apply to installations in certain premises, such as where there is public access, certain types of entertainment, housing for the aged or infirm, etc.

The impact of any Regulation which may bear on the function or cost of an installation should be brought to the attention of the purchaser at the earliest opportunity.

1.4   **Omission of protection 473**

Protection against either overload or fault current may be omitted where unexpected disconnection could cause danger. The designer should first consult

473-01-03
473-01-04
473-02-04

Regulations 473-01 and 473-02 to establish the conditions under which protection may be omitted.

The omission of overload or fault current protection, or a reduction in its effectiveness, can only be justified if danger is prevented by other means, or where the opening of the circuit would cause greater danger than the overload or fault condition.

This calls for a careful consideration of the balance of risks involved. The integrity of the equipment involved, including its ability to contain safely any possible effects of an overcurrent, such as arcs, hot particles, fire and dangerous fumes etc, should be examined.

The design should mitigate the effects of the omission of the overcurrent device e.g. by rating the circuit for any current reasonably likely to arise.

1.5 Protective devices 432

Protection may be provided by a single device which detects and interrupts both overload and fault current. Alternatively, separate devices may be used for each task. 432-02-01

There are special provisions for protection of motor starter circuits, see the guidance on Section 435 in paragraph 1.7.3.

1.6 Duration of overcurrent 431 434

Although BS 7671 does not impose a specific upper limit on the duration of overcurrent, except where there is a risk of shock or of damage to equipment, it is prudent to select circuit protection so that fault current does not persist any longer than is absolutely necessary. In a practical installation there is little or no control over the paths fault current may take, especially earth fault current, and danger or damage may occur in unexpected ways and to items not directly associated with the circuit.

Note that the data provided in Regulation 434-03-03 is valid for periods not exceeding 5 seconds and this tends to set an implied upper limit to fault current duration. 434-03-03

The duration of small overloads should be controlled by suitable attention to load assessment, see the guidance on Regulation 433-01-01 in paragraph 2.2.5.

1.7 **Co-ordination and discrimination**

Operation of protective devices in series

1.7.1

By reason of the method of power distribution adopted, there may well be two or more overcurrent protective devices involved with a given fault current or overload. For example, where a main device protects a feeder to a distribution board with protection for a number of outgoing circuits, or where protection against overload and fault current is provided by two separate devices. There are then two aspects which need careful consideration: discrimination between operating characteristics and co-ordination of their fault current breaking capacities and fault energy withstands.

1.7.2

Discrimination between devices

533

It is a matter of convenience and fitness for purpose of an installation that disconnection interrupts the faulty or overloaded circuit only. It can also be a matter of safety.

533-01-06

Correct selection and comparison of the characteristics of each device will ensure that only the device electrically nearest to the cause of the overcurrent operates. The breaking capacity of the downstream device must therefore be suitable for the highest prospective current at its point in the circuit.

The operating characteristics of a given device lie within a band, the lower boundary of which constitutes a specified non-fusing or non-tripping characteristic, the upper boundary a specified fusing or tripping characteristic. The manufacturer can provide guidance or information such that a desk comparison between the non-operating limit of the upstream device with the operating limit for the downstream device will normally be sufficient to select devices to give correct discrimination.

When the prospective current is high and interruption times are of the order of 1 or 2 half cycles it is unwise to attempt to assess discrimination by overlaying the characteristics. A combination of devices can behave very differently from when they are separate and advice should be sought from the manufacturers, who can recommend suitable combinations.

As a rule of thumb, satisfactory discrimination between two similar types of fuse link is usually achieved when the downstream link has a rating equal to about half that of the upstream one. For other devices a much higher ratio may be required, and in some circumstances may not be possible.

When discrimination is achieved, each protective device has to comply with the requirements of Regulations 432-02-01, 434-01-01 and 434-03-01 with regard to making and breaking prospective fault current at its point in the circuit.

432-02-01
434-01-01
434-03-01

1.7.3   *Co-ordination of devices*

434
435
473

It is sometimes not feasible or economical to provide a downstream device which has a sufficiently high fault breaking capacity.

431-01-02
435-01-01

A common example of this is the combination of a fuse or circuit-breaker with a motor starter. In general, motor starters are not designed, nor are they intended, to interrupt short-circuits or earth fault currents of a similar magnitude. However, in combination with the fault protection, they must be able to close on to the prospective fault current and to interrupt currents equal to the maximum overload capability of the associated motor. Fault protection for the starter circuit has to be provided by the fuses or circuit-breaker at the origin of the circuit, in compliance with Regulations 434-03-01 and 473-02-01.

434-03-01
473-02-01

The breaking capacity of the fault current protection device must be adequate to interrupt prospective fault currents, but its characteristics must also be such that it does not interrupt starting currents. Such characteristics are likely to be too high to operate on overload, either for the motor or to comply with Regulation 433-02-01 for the circuit conductors.

433-02-01

In motor circuit applications where circuit-breakers or gG type fuse links are used, the need to withstand motor starting currents usually dictates a higher current rating than would be selected on the basis of motor full load current.

However, to meet this special role, circuit-breakers providing fault protection only may be used and extended ratings of motor circuit protection fuse links with gM characteristics are available giving economies in both fusegear size and cost. These gM motor circuit fuse links have a dual rating: a maximum continuous rating based on the equipment in which they are fitted, for example a fuse carrier and base, and a rating related to the operational characteristics of the load.

In the type designation these two ratings are separated by the letter M. For example, 32M50 represents a maximum continuous rating of 32 A (governed by the associated fitment) and an operational characteristic of a 50 A fuse link.

Regulation 435-01-01 requires that the energy let-through of the fuse or circuit-breaker shall not exceed the withstand capability of the motor starter. In general, it is important that the overload relay does not initiate the opening of the starter contacts before the fault protection has had time to operate.

435-01-01

Where the fault protection is incorporated in one piece of equipment with the starter, it is the responsibility of the manufacturer to ensure that correct co-ordination is achieved. Where the devices are separate the manufacturer of the starter can provide guidance on the selection of a suitable fuse or circuit-breaker for fault current protection.

BS EN 60947-4-1 recognises the three types of co-ordination with corresponding levels of permissible damage to the starter. Co-ordination of Type C in BS 4941 or Type 2 in BS EN 60947-4 (i.e. no damage, including permanent alteration of the characteristics of the overload relay, except light contact burning and a risk of contact welding) is required if the starter is to continue to provide overload protection complying with Regulation 433-02-01.

433-02-01

Examination and maintenance of both devices is necessary after a fault.

The overload relay on the starter is arranged to operate for values of current from just above full load to the overload limit of the motor, but it has a time delay such that it does not respond to either starting currents or fault currents. This delay provides discrimination with the characteristics of the associated fuse or circuit-breaker at the origin of the circuit.

The starter overload relay can provide overload protection for the circuit in compliance with Regulation 433-02-01 on the basis that:

    (i) its nameplate full-load current rating or setting is taken as I_n, and

    (ii) the motor full-load current is taken as I_b, and

    (iii) the ultimate tripping current of the overload relay is taken as I_2.

Where the overload relay has a range of settings then items (i) and (iii) should be based on the highest current setting, unless the settings cannot be changed without the use of a tool.

It is unlikely that the overload withstand of the circuit conductors will coincide with that of the motor, so that overload settings should be determined by the lower requirement. Alternatively, an additional trip system, based on detection of motor temperature may be used.

As a starter provides overload protection only, Regulation 473-01-02 permits it to be located anywhere along the run of the circuit from the distribution board providing there is no other branch circuit or outlet between the starter and the motor.

1.7.4

Back-up co-ordination

434

Another area where co-ordination is needed occurs when it is necessary to provide back-up for a fault current protective device with another one upstream of higher rating or higher fault current breaking capacity. It is not always satisfactory to evaluate the performance of the two devices by a comparison of their characteristics. For example, with short

433-02-01

473-01-02

434-03-01

operating times such an approach would not take account of the effect of either device on the magnitude of the fault current and their performance in series may be different from their individual performances.

Information should be obtained from the manufacturer of the downstream device as to the type of upstream fuse or circuit-breaker best suited to the prospective fault current.

Section 2 — Protection Against Overload

2.1 General 433

An overload is a current the value of which exceeds the rated current of an electrically sound circuit. It may be caused by a user deliberately or accidentally using more power than the circuit is intended for, or by a fault in equipment supplied by the circuit, or it may be caused by a characteristic of the load. It is not due to an electrical fault in the circuit itself.

Part 2

Damage by overload can take the form of accelerated deterioration of insulation, contacts and connections and deformation of some materials, thereby reducing the safe life and capability of equipment.

An overload caused by plugging in too many appliances may not be large, but it could last long enough to cause prolonged excessive heating of the conductors and probably of some of the accessories.

On the other hand, an overload caused by starting a motor may be several times the circuit rating, but its duration is not long enough to cause unacceptable overheating. However, where frequent motor starting or reversing at short intervals is expected, heating will be cumulative and the circuit components should be suitable for the higher duty.

The important characteristics of an overload are therefore the magnitude of the current and its duration.

2.2 Load assessment 433

Proper assessment of the load current I_b is essential for the safe application of these regulations.

433-01-01

This is reflected in the fundamental principles (Chapter 13) and includes consideration of the maximum demand on each distribution and final circuit.

132-01-04

Although overload protection is based on conductor temperature, it is equally important that all joints and connections are also capable of withstanding the currents and consequent temperatures expected. 131-06-01
131-07-01

When the circuit supplies a single unvarying load or a number of such loads, I_b is obtained by simply summing all the loads concerned. In case of doubt, or for simple installations, this is the safest course.

2.2.1 *Diversity*

311 In some cases it may be appropriate to recognize that not all the loads are operating at the same time and to apply a diversity factor. Determination of a diversity factor requires care and experience and account should be taken as far as possible of future expansion of the load or of change of duty. 311-01-01

Further advice on the determination of diversity factors is given in Guidance Note 1, Selection & Erection, Appendix H. GN1

2.2.2 *Varying loads*

533 Where a load changes in a known manner it is possible to equate the design current I_b to a thermally equivalent constant load. Varying loads include both irregular pulses of load and regular or cyclic variations. 533-02-01

2.2.3 *Thermally equivalent current*

433 Where a circuit supplies a number of independent loads, or a fluctuating load, a simple procedure is to assess the value of the highest total load on the circuit and to substitute this for I_b in Regulation 433-02-01. 433-02-01

This approach is adequate for circuits supplying small loads where further design effort is not justified.

However, Regulation 533-02-01 indicates that regular repetitive loads (cyclic loads) should be assessed on the basis of a thermally equivalent constant load. 533-02-01

The benefits of such an approach can be extended to any varying load.

Descriptions of methods to calculate thermally equivalent loads are outside the scope of this Guidance Note. Such calculations can be tedious and it may be helpful to use the following routine, which will indicate whether there is anything to be gained by such a calculation.

2.2.4   *Test for thermally equivalent load*

(i) Include in the total load all contributions whatever their magnitude and duration. Initially, make I_b equal to the maximum current, I_{max} (see Fig 1)

(ii) Select a conductor size to fulfil the condition:
$I_b \leq I_n \leq I_z$
See Regulation 433-02-01 and Appendix 4 of BS 7671.

Fig 1: **Illustrating a varying load**

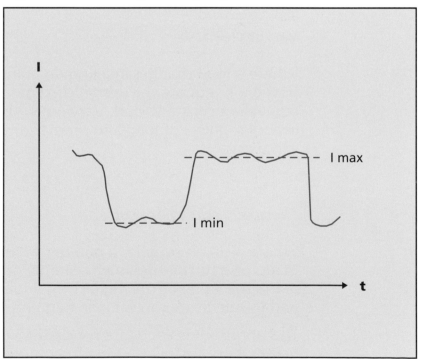

(iii) If the duration of the peak current, I_{max}, is greater than the duration given in Table 1 for the conductor size and type of cable selected, no advantage can be gained by calculating the thermally equivalent current and step (ii) above will give the correct cable size. This applies to a single pulse of load or to a repetitive load

(iv) If the duration of I_{max} is less than the duration in Table 1, it may be possible to select a smaller conductor size. This will depend on the value of the preceding lower current and its duration

(v) Calculate the ratio $X=$(duration of I_{max})/(duration in Table 1) for the size of cable selected in step (ii). From Fig 2 read off the appropriate value of Y

(vi) If the duration of the lower current, I_{min}, is less than Y times the duration of I_{max} no worthwhile reduction in conductor size will be possible. Step (ii) will give the correct size of conductor.

TABLE 1
Duration of load current for equivalent constant load (Duration in minutes)

Conductor size mm^2	1-core cable	2-core cable	3-core cable	4-core cable
1.0	—	12	11	12
1.5	—	13	12	15
2.5	—	15	14	16
4	—	16	16	20
6	—	18	19	24
10	—	22	21	26
16	—	26	27	33
25	—	35	39	49
35	—	41	46	58
50	36	37	47	62
70	42	46	56	74
95	50	55	67	89
120	58	62	77	102
150	67	73	90	120
185	78	84	103	136
240	92	102	124	165
300	106	118	143	192
400	126	143	170	227
500	157	—	—	—
630	172	—	—	—

Fig 2: **Test for variable load**

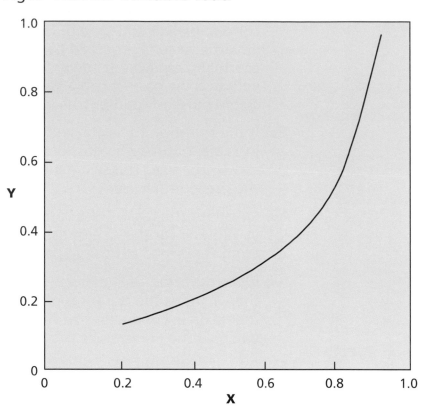

(vii) If the duration of I_{min} is greater than Y times the duration of I_{max}, calculation of the thermally equivalent current (which will be less than I_{max}) may lead to lower values for I_b and I_n and possibly a smaller cable size. The thermally equivalent current, and hence any benefit by way of a smaller conductor size, will depend on the value of I_{min}.

2.2.5   *Small overloads of long duration*

433   If I_b is equal to the sum of all expected loads, or to a thermally equivalent load, there should be no overload which will result in the conductor temperature exceeding its permitted working value. Overload protection then fulfils its proper function of protecting the circuit against events over which the designer has no control.

433-01-01

523-01
Table 52B

However, examination of the characteristics of many overload protective devices will reveal that, for practical reasons, they may not provide overload protection against currents just exceeding the rating of the associated conductors and equipment. For this reason it is important that every care is taken to avoid even small overloads of a protracted nature. (Small overloads are further discussed in 2.3.6)

Ageing and deterioration of insulation and connections increase rapidly at temperatures exceeding the rated values. The limits permitted assume that excess temperatures will be very infrequent and generally due to unforeseen situations. It is not practicable to place a general limit on the permissible number of overload events.

To avoid deterioration, it is essential that the design load of a circuit is high enough to include all foreseeable peak loads of a protracted nature.

It is never acceptable to use a fuse or a circuit-breaker as a load limiting device.

2.3 Selection of protective device and conductor size 433

There are two regulations which need to be considered when selecting an overcurrent device.

The first specifies certain relationships between:

433-02-01

I_b the load (design) current, and

I_n the nominal current or current setting of the protective device, and

I_z the current-carrying capacity (rating) of the circuit conductors, and

I_2 the current causing effective operation of the device.

The relationships between I_b, I_n, I_z and I_2 can be set out as equations, or in diagrammatic form as in Fig 3.

(i) $I_b \leq I_n$

(ii) $I_n \leq I_z$

(iii) $I_2 \leq 1.45 \times I_z$.

Note that the load current, I_b, is the starting point of this regulation.

Fig 3: **Overload protection**

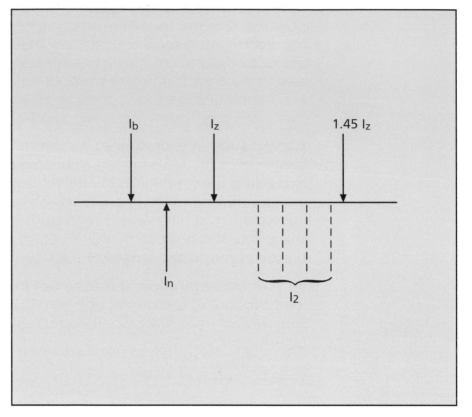

Conditions (i) and (ii) provide the basic relationships for the selection of a suitable size or setting of the overload device ($I_b \leq I_n$), and an adequate conductor ($I_n \leq I_z$). The significance of selecting the conductor rating in this way is that it provides a link between the operating characteristic of the device and the conductor rating and paves the way for condition (iii).

The alternative term 'current setting' for I_n applies to devices which have adjustable characteristics, when the nominal size might be larger than the current-carrying capacity of the conductor.

Where the device is adjustable, the means of adjustment should be sealed, or constructed so that alteration cannot be made without the use of a tool. A non-adjustable device should be used where untrained personnel may have access.

Condition (iii) provides the overload protection required by Regulation 433-01-01, but note that it is concerned primarily with the conductors. Protection for electrical equipment in the circuit, including the overload device itself, should be covered by careful selection of equipment, made to a standard quoted in the Regulations or to an equivalent.

433-01-01

The factor 1.45 in condition (iii) is based on a combination of experience and investigation. This has shown that the types of cable considered in BS 7671 can safely withstand a small but undefined number of periods at excess temperatures corresponding to currents not greater than 1.45 times their current-carrying capacity. However, such currents must not persist for long periods.

The second regulation always to be considered with Regulation 433-02-01 is 433-02-02. This states that, for the devices specified in Regulation 433-02-02 (these are the devices commonly used in the UK with the particular exception of rewireable fuses), the requirement for $I_z \leq 1.45\, I_2$ will be met if $I_n \leq I_z$.

433-02-02

2.3.1   *Selection of conductor size, and value of I_z*

433   Values for I_z, the current-carrying capacity of conductors, can be obtained from Appendix 4 of BS 7671. Attention is drawn to Sections 4, 5 and 6 of that Appendix where the selection of a conductor size to satisfy Regulation 433-02-01 is described.

433-02-01
Appx 4

Only the magnitude of the load current is specified in Regulation 433-02-01, the duration is taken care of by using a protective device having a suitable time/current characteristic.

433-02-01

A device complying with one of the British Standards listed in Regulation 433-02-02 has a characteristic designed so as to break an overload current within the limit of 1.45 times the conductor rating as required by condition (iii), provided that it has been chosen so as to comply with condition (ii).

433-02-02

Furthermore, its characteristic is such that the duration of an overload will be limited to a safe value.

Where fuses are used it is important to note that only the gG type fuse to BS 88-2.1 will automatically provide overload protection complying with both conditions (ii) and (iii) of Regulation 433-02-01. A 'gG' type fuse refers to a current limiting fuse link which is intended for general purpose use, particularly where overload protection is required. It should be distinguished from types 'gM' (for motor circuits) and 'a' (partial-range breaking capacity) fuse links, which are not intended for overload protection.

433-02-01

Regulation 433-02-03 applies when semi-enclosed fuses to BS 3036 are used because they operate when the overload current is twice their nominal current.

433-02-03

Correct protection can be obtained by modifying condition (ii) so that the nominal current of such a fuse does not exceed 1.45/2 = 0.725 times the rating of the associated conductor. The practical effect of this modification is to increase the size of conductor required for a given load.

Further comments on the use of BS 3036 fuses are made in Appendix 4 of BS 7671.

Appx 4

2.3.2

Nuisance tripping of overload protection

Unintentional operation of overload devices may occur when the load current contains peaks lasting up to a few seconds. For example, motor starting currents and inrush currents due to energising transformers or large tungsten lamp loads, where selection of the conductor size has been made by putting I_b equal to the normal maximum running current. Such peaks are unlikely to result in unacceptable conductor heating, unless, in the case of a motor, the starting period is unusually long (for example, due to a high inertia load).

However, unnecessary operation may cause inconvenience or danger. A different size or type of device may be needed whose characteristics provide reduced sensitivity or a time delay for short durations together with protection complying with Regulation 433-02 for long duration overloads.

433-02

For motors larger than the fractional kilowatt sizes the problem is dealt with by providing a suitable delay to the overload trips in the starter. (See the description of this subject in paragraph 1.7 of this Guidance Note.)

Different types of circuit-breaker (cb) to BS EN 60898 (which has replaced BS 3871) are available which operate at various values of short duration overcurrent (instantaneous trip current (I_n)) as shown in Table 2.

The different types of circuit-breaker available can be split into two categories.

Circuit-breakers for household and similar applications to BS EN 60898 (formerly known as mcbs) and rcbos to BS EN 61009-1.
There is a wide range of circuit-breaker characteristics that have been classified according to their instantaneous trip performance and Table 2 gives some information on the application of the various types available. These limits are the maximum allowed for miniature circuit-breakers to BS 3871 (now withdrawn) and circuit-breakers to BS EN 60898-1, but it should be noted that manufacturers may provide closer limits. Note: The same types B, C and D also apply to rcbos to BS EN 61009-1.

Low voltage switchgear and controlgear Part 2 circuit-breakers to BS EN 60947-2.
Unlike circuit-breakers to BS EN 60898-1, circuit-breakers complying with BS EN 60947-2 do not have defined characteristics and manufacturers' data must be used. BS EN 60947-2 Amendment 2 : 2001 now includes Annex L, which is for circuit-breakers not fulfilling the requirements for overcurrent protection (cbis) derived from the equivalent circuit-breaker. A class X cbi is fitted with integral short-circuit protection, which may, on the basis of the manufacturer's data, be used in conjunction with the starter overload relay, for short-circuit protection.

TABLE 2
Selection of cb or rcbo type

Type	Instantaneous trip current	Application
1*	2.7 to 4 I_n	Domestic and commercial installations
B	3 to 5 I_n	having little or no switching surge
2*	4 to 7 I_n	General use in commercial/industrial
C	5 to 10 I_n	installations where the use of fluorescent
3*	7 to 10 I_n	lighting, small motors etc can produce switching surges that could cause unwanted tripping of a Type 1 or B cb. Type C or 3 may be necessary in highly inductive circuits such as banks of fluorescent lighting
4*	10 to 50 I_n	suitable for transformers X-ray
D	10 to 20 I_n	machines, industrial welding equipment etc where high inrush currents may occur

* Types 1, 2, 3 and 4 are no longer available, but have been shown in the table because these are often still found in existing installations.

Table 2 also provides guidance on the selection of cb type for a given application.

Again, reference to manufacturers' data is advisable.

For guidance on the selection of the type and number of loads that can be simultaneously switched reference to manufacturers' data is recommended.

552   Fig 4: **Typical star-delta arrangement**

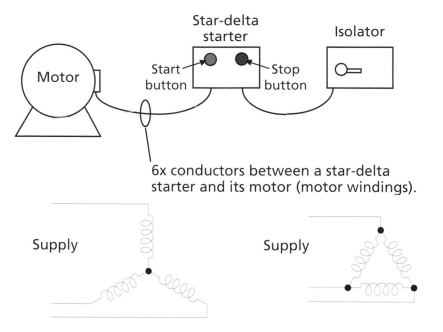

6x conductors between a star-delta starter and its motor (motor windings).

Motor windings connected in star for start.

Motor windings connected in delta for run.

When a motor is controlled by a star-delta starter the ends of the three motor windings are extended from the motor terminals to the starter. The connections of the windings in star for start, and delta for run are made at the starter.

The steady running current in each of the six conductors between a star-delta starter and its motor is 58 per cent of that through the conductors supplying the starter. However, motor cables are usually run as a group having six conductors so that their current-carrying capacity is 80 per cent of a single circuit three conductor capacity.

552-01

It follows that the single circuit rating of motor cables should be at least 72 per cent (0.58/0.80) of that of associated supply cables. Examination of rating tables shows that the empirical rule by which motor cables are half the cross-sectional area of the supply cable does not always meet this requirement. Neither does it take into consideration possible differences in cable type or installation conditions.

A simple procedure for the correct selection of motor cables, whereby the motor cables are chosen to have a rating, I_z, of at least 72 per cent of that of the supply cables, is illustrated by the following example, which uses the symbols and Tables of Appendix 4 of BS 7671.

Appx 4

Assume a full-load motor current of 37 A = I_b.

Supply cable, to be three-core thermoplastic (pvc) swa on a perforated tray. See Table 4D4A of BS 7671.

Table 4D4A

Ambient temperature 30 °C (C_a, C_g, C_i = 1).

See Appendix 4, Section 6, of BS 7671.

Appx 4

Motor cables, to be single-core 70 °C thermoplastic (pvc) in conduit. See Table 4D1A.

Table 4D1A

Ambient temperature 40 °C (C_a = 0.87). See Table 4C1 of BS 7671.

Table 4C1

For the supply cable, Regulation 433-02-01 ($I_b \leq I_n$) is met by a starter for which I_n = 40 A.

433-02-01

The cable size is then selected as in Appendix 4, 6.1.1, of BS 7671.

Appx 4

$$I_t \geq \frac{I_n}{C_a . C_g . C_i} = \frac{40}{1 \times 1 \times 1} = 40 \text{ A.}$$

A three-core 6 mm^2 cable has a tabulated rating, I_t = 45 A. (Table 4D4A, column 5)

Table 4D4A

If the delayed overload trips operate at, say, 50 A, condition (iii) of Regulation 433-02-01,

433-02-01

$I_2 \leq 1.45 \, I_z$ where $I_z = C_a . C_g . C_i . I_t$

is met because I_2 = 50 A, and

1.45 x 1 x 1 x 1 x 45 = 65 A.

For the motor cables, the procedure requires that

$$I_t \geq \frac{I_n}{C_a . C_g . C_i} \times 0.72 = \frac{40}{0.87 \times 1 \times 1} \times 0.72 = 33 \text{ A.}$$

6 mm^2 single-core cables in conduit, which have a single circuit tabulated rating, I_t, of 36 A will be suitable (see Table 4D1A, Column 5, of BS 7671).

Table 4D1A

If overload protection is satisfactory for the supply cables, the motor cables will also be protected. This can be demonstrated as follows:

I_2 = 50 (which is a line current)

For the motor cables, which are a double circuit in conduit,

$$I_z = I_t.C_a.C_g.C_i$$

$$= 36 \times 0.87 \times 0.8 \times 1 = 25 \text{ A}$$

$$1.45 \, I_z = 1.45 \times 25 = 36.25 \text{ A}.$$

The line equivalent of this current in the delta connected motor cables is:

$$\sqrt{3} \times 36.25 = 63 \text{ A}.$$

In this example, motor cables having a cross-sectional area (csa) of 50 per cent of the supply cable would be clearly inadequate. This is partly due to the use, in this example, of different types of cable and a higher ambient for the motor cables.

Even if the example had specified cables enclosed in conduit and the same ambient temperature for both runs of cable, the 50 per cent csa rule would fail, being saved only by the need to round up the size of the motor cables to the nearest larger standard size.

For higher loads and larger cables this rounding up is not always necessary and the 50 per cent rule is inadequate.

2.3.4

Cables in Parallel

433

It is permissible to use only one device to provide overload protection for circuits composed of conductors in parallel, provided that the device will operate before damage occurs to any of the cables. 433-03-01

To achieve this it is necessary that the load current divides equally between the conductors.

As a first step this can be achieved by using conductors of the same length, cross-sectional area and material. It is then assumed that overload currents divide equally between the conductors and that I_z is the simple sum of the current-carrying capacities of the conductors. There are some further requirements, however, dependent on the type of cables involved, as the arrangement of the cables may introduce mutual heating between them and reduce their current-carrying capacities.

Single-core cables

For single-core cables, only certain cable arrangements provide reasonable current sharing. Examples are given in Fig 4. Other arrangements may provide acceptable current sharing, but this should be checked

Multicore cables

For multicore cables, the conductor arrangements can be of two types:

(a)   all the phase (and neutral) lines are carried in each cable; a circuit protective conductor is either included in each cable (in armoured cables it would be the armour) or, if the cables are not armoured, it may be run as a separate conductor.

Satisfactory current sharing is easily achieved with this arrangement provided that cables are of the same size, type and length. The total current-carrying capacity is proportional to the number of cables, provided that any reduction in rating due to grouping is taken into account

(Note, even though the cables in parallel are in effect one circuit, the cables still need to be treated as though each was a separate circuit from a grouping point of view.)

Refer to Appendix 4, Table 4B1 in BS 7671 : 2001.   Table 4B1

(b)   for unarmoured cables only it is feasible to use them as single-core cables by connecting all live conductors together. However, any circuit protective conductor should be a separate conductor and not a conductor in one or more of the cables. In the event of a cable fault the fault current withstand of such a conductor would be inadequate. Further, currents induced in loops formed by such conductors will reduce the current-carrying capacity of the cables.

It is important to bear in mind that the total current-carrying capacity of cores paralleled in this way is not proportional to the number of cores. Unless better information is available it is prudent to expect four-core cables to carry only three times the rating of one core.

Fig 5: **Cable configurations for single-core cables in parallel**

2 CABLES PER PHASE (equal current sharing)

(N) = 1 neutral, (N)(N) = 2 neutrals, phase cables equally spaced, or touching. Trefoil groups touching.

---

3 CABLES PER PHASE (no complete sharing possible with any configuration,
example gives about 5 per cent unbalance,
and up to 10 per cent circulating current in neutrals)

(*) = clearance De   (Where De = the outside diameter of the cable.)

(*) = clearance 3 De   (Where 3 De = 3 times the outside diameter of the cable.)

1 neutral may be run in middle of each inter-group spacing.

---

4 CABLES PER PHASE (flat formations provide equal sharing, no circulating
current in neutrals, trefoils as for 3 cables per phase)

Cables touching or equally spaced

(N) = 1 neutral, up to 4 positions may be used

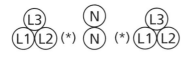

(N) = 1 neutral, up to 4 positions
may be used
If position is not used space
must be maintained

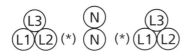

Trefoils touching
Vertical clearance
between trefoils 2 De

(*) = clearance De   (Where De = the outside diameter of the cable.)

Fig 6: **Typical ring final circuit arrangement**

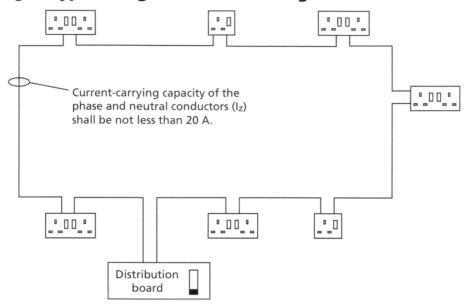

Current-carrying capacity of the
phase and neutral conductors (I_z)
shall be not less than 20 A.

Distribution
board

Regulation 433-02-04, as amended by Amendment No
1 February 2002, sets out the requirements for ring
circuits. This regulation requires that the current-
carrying capacity (I_z) of the cable is not less than 20A,
and that, under the intended conditions of use, the
load current in any part of the ring is unlikely to
exceed for long periods the current-carrying capacity
of the cable. This is reinforced by Regulation 433-01-01,
which requires that every circuit be designed 'so that
a small overload of long duration is unlikely to occur'.

433-02-04

433-01-01

This last point is important because a small overload
current may not be sufficient to operate the
protective device, but could raise the temperature of
the conductors above the rated value and damage the
cable. An explanation of what is a small overload and
what is a long period is given in paragraph 2.3.6.

Ring circuits are intended to provide large numbers of
conveniently placed outlets, rather than a high load
capability. The limitation on the floor area served by a
ring circuit in household installations is intended to
provide a simple means of load limitation.

For household installations, a single 30 A or 32 A ring
final circuit may serve a floor area of up to 100 m².
However, careful consideration should be given to the

loading in kitchens, which may require a separate circuit. Socket-outlets for washing machines, tumble dryers and dishwashers should be located so as to provide reasonable sharing of the load in each leg of the ring, or consideration should be given to a suitable radial circuit.

Immersion heaters fitted to storage vessels in excess of 15 litres capacity, or permanently connected heating appliances forming part of a comprehensive electric space heating installation, should be supplied by their own separate circuit.

Table 4D5A of Amendment No 1 of BS 7671:2001 gives the current-carrying capacity of 70 °C thermoplastic (pvc) insulated and sheathed flat cable with protective conductor which is often used in household installations.

For other premises or uses, a careful assessment should be made of the maximum load, its time diversity and its distribution around the ring, which may not be related to floor area. This is essential for non-domestic use because the circuit protection must not be used as a load limiter.

2.3.6   *Load assessment*

What is a small overload?

On the question of what is a small overload, guidance can be taken from the BS 88 fuse standard and the BS EN 60898 and BS EN 61009-1 circuit-breaker standards. For a BS 88 fuse the 'non-fusing' current is 1.25 times the fuse rating and for BS EN 60898 circuit-breakers and BS EN 61009-1 RCBOs the 'non-tripping' current is 1.13 times the rating of the circuit-breakers. (The non-fusing and non-tripping currents are the currents that the fuse or circuit-breaker must carry for the conventional time without operating.) The fusing current is 1.6 times the fuse rating and the tripping current is 1.45 times the circuit-breaker rating. (These are the currents at which the fuse or circuit-breaker must operate within the conventional time, which is 1 hour for fuses and circuit-breakers rated up to and including 63 A.) For the tests, the fuse starts from cold, the circuit-breaker starts from a heated condition.

There is also a test under different conditions where the fuse must operate within the conventional time at 1.45 times its rating. BS 7671 accepts the 1.45 factor and hence accepts that an overload can be tolerated for the conventional time, that is, 1 hour for a 32 A fuse.

433-02-01

Circuit-breakers to the standards above must operate within the conventional time (i.e. within 1 hour) at 1.45 times their rating.

Therefore, it seems reasonable to suggest that a small overload is between 1.25 and 1.45 times the rating of the overload protective device.

What is a long period?

A long period cannot be less than the 'conventional operating time' (one hour) of a fuse because that is already allowed. It is also worth noting that the time/current characteristic curves in Appendix 3 of BS 7671 for a 32 A BS 88 fuse stop at around one hour. It is reasonable to suggest that a 'long period' is a time greater than that for which information on fuse or circuit-breaker performance, that is, one hour.

2.3.7

Mineral insulated cables and ring circuits

Regulation 433-02-04 as amended gives a minimum cross-sectional area of both phase and neutral conductors of 2.5 mm², except for two-core mineral insulated cables to BS 6207, for which the minimum is 1.5 mm². There has been concern the 73 °C operating temperature of mineral cables (for a 70 °C sheath temperature) might exceed that allowed for accessories by the appropriate standard (a temperature of 70 °C is assumed in the type tests of the standards). The current-carrying capacity of two-core 1.5 mm² light duty mineral insulated cable is 23 A (see column 2 of Table 4J1A of BS 7671). At this loading the estimated conductor temperature is 73 °C (sheath temperature 70 °C) which would overheat the accessory. This resulted in concerns regarding the advisability of applying the 20 A relaxation of Regulation 433-02-04.

433-02-04

Table 4J1A

However, assuming a maximum current of 20 A and that temperature rise is proportional to the square of the current, then temperature rise for a ring circuit with a maximum current of 20 A is given by:

$$= \left(\frac{20}{23}\right) \times (73 - 30)\ °C$$

$$= 32.51\ °C$$

Assuming an ambient temperature of 30 °C gives a conductor temperature of 62.51 °C which is below 70 °C.

2.4 Omission of protection against overload current 473

Overload current protection may be omitted altogether under certain circumstances. However, it should not be omitted, or its sensitivity reduced, without very careful consideration; see comments on Regulations 473-01 and 473-02 in paragraph 1.4 of this Guidance Note.

Regulation 473-01-03 allows the omission of devices for protection against overload for circuits supplying current-using equipment where unexpected disconnection of the circuit could cause danger. The regulation gives the following examples:

473-01-03

In such situations consideration shall be given to the provision of an overload alarm.

- Exciter circuits of rotating machines.

- Circuits which supply fire extinguishing devices. For example, a sprinkler system.

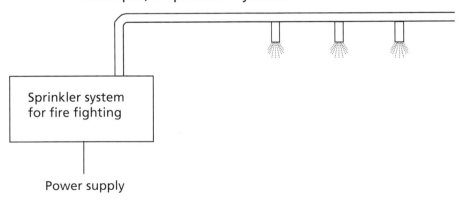

Sprinkler system for fire fighting

Power supply

- Supply circuits of lifting magnets.

Electro-magnets are used in, e.g. scrap yards to lift and carry loads. If such a magnet is de-energised while in operation this could cause damage or injury.

- Secondary circuits of current transformers.

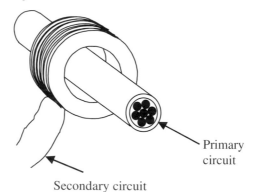

Primary circuit

Secondary circuit

The secondary circuit of a current transformer must be kept closed when current is flowing in the primary winding. If not, the peak value of emf induced in the secondary circuit will be many times the nominal value, and could be large enough to cause breakdown of the insulation and danger.

Regulation 473-01-04 gives a relaxation where devices for protection against overload need not be provided.

473-01-04

These are:

- For a conductor situated on the load side of the point where a reduction occurs in the value of current-carrying capacity, where the conductor is effectively protected against overload by a protective device placed on the supply side of that point.

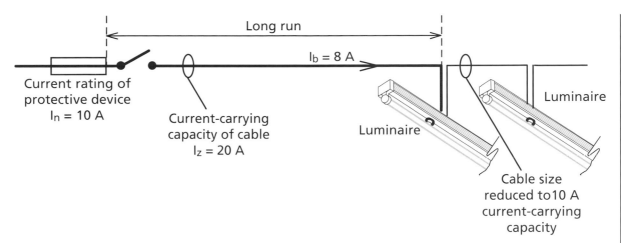

Long run

$I_b = 8\ A$

Current rating of protective device $I_n = 10\ A$

Current-carrying capacity of cable $I_z = 20\ A$

Luminaire

Luminaire

Cable size reduced to 10 A current-carrying capacity

The diagram above shows a circuit in which a larger cable than necessary (from a current-carrying point of view) is used to reduce voltage drop, where the route to the first luminaire from the distribution board is a long run. The cable from the first luminaire supplying the remainder of the luminaires on the circuit is then reduced, but is protected against overload by the 10 A device.

- For a conductor which, because of the characteristics of the load or supply, is not likely to carry overload current.

- At the origin of an installation where the distributor provides an overload device and agrees that the device affords protection to the parts of the installation between the origin and the main distribution point of the installation where further overload protection is provided.

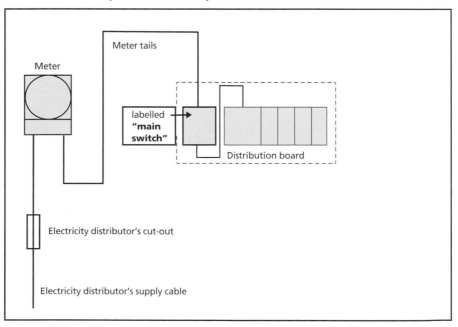

Meter tails

Meter

labelled **"main switch"**

Distribution board

Electricity distributor's cut-out

Electricity distributor's supply cable

Section 3 — Protection Against Fault Current

3.1 Types of fault to be considered 130 434

The Regulations require that each circuit should be provided with a device capable of breaking any fault current flowing in that circuit before it causes danger due to thermal or mechanical effects.

130-05
130-04
434-01

Fault currents result from an insulation failure or the bridging of insulation by a conducting item.

There are two types of fault:

— a short-circuit, where the current flows between live parts (which may include the neutral)

— an earth fault, where the current flows between a live part and earth. The latter includes any conductive part or conductor which is connected to the earth terminal or is substantially in contact with the mass of earth.

The types of damage envisaged include:

— overheating and mechanical damage to conductors, connections and contacts

— deformation and deterioration of insulation leading to future electrical breakdown

— risk of ignition of materials adjacent to a conductor, due to its high temperature or to prolonged arcing at the fault or at a break in the circuit caused by the fault current.

3.2 Nature of damage and installation precautions 434 543

In general, fault currents are much larger than either load currents or overloads and the amount of damage which can be done is so great that rapid interruption is essential.

BS 7671 requires that fault energy be limited to a value which will not cause a temperature rise sufficient to damage insulation (see Regulations 434-03-03 and 543-01-03) and appropriate temperature limits are given for live conductors in Table 43A and for protective conductors in Tables 54B to 54E. Table 54F, for bare conductors, is based on general considerations with regard to damage to the conductor or to unspecified materials which may be in contact or near by. Damage means excessive deformation or reduction in thickness of the insulation.

434-03-03
543-01-03
Table 43A
Table 54B
Table 54C
Table 54D
Table 54E
Table 54F

The temperature limits given in these tables are based on the physical properties of the insulating materials, together with results of tests made on cables supported so that mechanical forces on the insulation would be representative of the values likely to occur in practice.

3.2.1   *Installation precautions*

Poor installation methods or adverse installation conditions may result in the imposition of greater forces, with the risk that damage may be excessive. For this reason good workmanship is essential when installing cables and methods of support should comply with those given in Chapter 52 of BS 7671.

Chap 52

In particular, careful attention should be paid to the minimum installation radii recommended in appropriate cable standards, the avoidance of excessive force during installation, contact with sharp edges, overtight supporting clamps, unsupported lengths, leaving a cable in tension and to all situations where local stressing might occur.

3.2.2   *Expansion forces*

All conductors expand lengthways when heated and cable support should be such that this expansion can be accommodated as uniformly as possible along the route. Supports and cable ties should be uniformly spaced and, in the case of larger cables, these should be set in a slightly sinuous path so that a small lateral movement is possible. Otherwise, expansion may accumulate at one point, causing severe bending, or at the ends where it may overstress the terminations.

Larger sizes of conductor, roughly 50 mm^2 or more, are capable of exerting considerable longitudinal expansion forces and, unless the expansion is accommodated uniformly along the route, these forces may cause serious distortion of the conductor tails at terminations or mechanical overstressing of the connections.

Local restriction of larger cables at bends, particularly at the smaller radius bends, will produce high local stresses on both insulation and sheath, with an increased risk of damage.

Cables buried directly in the ground have no opportunity for expansion relief by lateral movement and expansion forces are applied directly to joints and terminations. Cable and joint makers should be consulted to confirm that the installation can withstand the short-circuit temperatures envisaged.

Allowance must be made for the thermal expansion and contraction of long runs of steel or plastic conduit or trunking, and adequate cable slack provided to allow free movement. The expansion or contraction of plastic conduit or trunking is greater than that of steel for the same temperature change.

For busbar trunking systems BS EN 60439-2 identifies a busbar trunking unit for building movements. Such a unit allows for the movement of a building due to thermal expansion and contraction. Reference should be made to the manufacturer since requirements may differ with respect to design, current rating or orientation of the busbar trunking system (i.e. riser or horizontal distribution).

Cables crossing a building expansion joint should be installed with adequate slack to allow movement, and a gap left in any supporting tray or steelwork. A flexible joint should be provided in conduit or trunking systems.

3.2.3   *Electromagnetic forces*

434   Compliance with the temperature limits in Table 43A may not avoid the risk of damage due to electromagnetic forces.   434-01-01
Table 43A

These forces are unlikely to be important for circuits where fault current protection complying with Regulation 434-03-02 and rated at not more than 100 A is provided, but can become increasingly significant as fault currents increase above about 15 kA rms or about 20 kA pk. It should be borne in mind that an initial peak current approaching 1.8 times the rms value is possible when a fault is electrically near to a large transformer or generator source.   434-03-02

The cut-off characteristics of current limiting fault protection devices will act to restrict such high initial peak currents and the forces they would produce. Information can be obtained from the manufacturer of the device.

It can be assumed that *unarmoured* multicore cables of all sizes are strong enough to withstand electromagnetic bursting forces (i.e. mechanical forces, forcing the conductors apart) where protection is co-ordinated with cable size in compliance with Regulation 434-03-02 or with Regulation 434-03-03 when the fault current protection device is of the current limiting type. Cable makers should be consulted where let-through currents are likely to be in the region of 25 kA rms or higher.   434-03-02
434-03-03

It can be safely assumed that all *armoured* multicore extruded insulation cables are strong enough to withstand bursting forces without damage, whatever type of protective device is employed. However, where currents may exceed 25 kA rms, it is wise to consult the cable maker.

In the case of single-core cables, electromagnetic forces act to move the cables apart. The effect is dependent on cable size, type and spacing (the closer the cables, the greater the forces), on support provided by fixings and on the peak value of the current. Fault currents well in excess of 20 kA pk may result in severe bending of the cables and with higher currents there is the possibility of broken fixings. To reduce the risk of damage, cables should be strongly bound together at frequent intervals to resist electromagnetic forces, but should be supported or fixed at greater intervals to allow lateral movement and uniform relief of longitudinal expansion by snaking. Again, the cable maker can provide information on the type of binding and supports to use.

A weak point as regards both longitudinal expansion forces and electromagnetic forces is unsupported lengths of core at joints and terminations. Electromagnetic forces are highest in the region of the crutch of multicore cable terminations, i.e. where the conductors are closest. Adequate support, which may be in the form of a suitable gland or binding, to reinforce the end of the cable envelope and to support the cores at the crutch is desirable. Lateral support for cores should be considered where tails are long enough to bend and impose strain on terminals.

Adequate support must be given to cables in busbar chambers to prevent abrasion of insulation by electromagnetic movement.

| 3.3 | **Fault impedance and breaking capacity of protective device 131 432 434** | It is conventionally assumed that the impedance of the actual fault is zero. This makes calculation of the value of fault current straightforward, because only the impedances of the circuit conductors need be taken into account. Such an assumption neglects any arc voltage and provides the highest expected value of fault current at a particular point in a circuit, referred to as the prospective fault current, and hence the maximum current breaking capacity required of the device to be selected to interrupt the current. | 131-08-01 432-02-01 432-04-01 434-02-01 |

It is important for this selection to be correct, not only because the successful interruption of the current avoids unnecessary risk of damage, which in higher current circuits can be very serious, but also because, in the extreme, if the device fails to interrupt the current it is likely to be destroyed and the duty of interruption passes to the next device upstream. This next device is usually chosen to detect a much higher magnitude of fault current and may not operate, or may only operate after considerable damage and risk of danger to persons or property has occurred.

3.4 Position of fault current protection and assessment of prospective current 473

With certain exceptions, fault current protection is required at the origin of each circuit or, more generally, where there is a reduction in the size of conductor and hence in fault current withstand capacity.

473-02-01

3.4.1

Assessment of fault current

533

At the origin of a circuit, or point of reduction in conductor size, it is necessary to calculate both the highest and the lowest values of prospective fault current. The operating time and energy let-through ($I^2 t$) of a protective device depends on the current and it is necessary to check that conductors being protected can withstand the higher of the two $I^2 t$ let-through values. It is a requirement that the device is capable of safely interrupting the highest prospective current and that it will operate correctly for the lowest prospective fault current.

533-03-01

The highest value of prospective fault current is calculated assuming a fault immediately on the load side of the device. The impedances concerned are those upstream of the device, including that of the supply to the installation, and should include, if significant, the impedance of the device itself.

The lowest value of prospective fault current is calculated on the assumption that a fault occurs at the load end of the circuit or at the input to the next downstream fault current protective device.

In a single-phase circuit the line to earth prospective fault current may be greater than that for a line to neutral fault.

3.4.2

Fault current and impedance at origin of installation

313

BS 7671 requires that the prospective short-circuit current at the origin of the installation be ascertained.

313-01-01

The Electricity Association have indicated in Engineering Recommendation P25/1 that at a service tee-off of a public supply network a typical maximum value of 16 kA may be assumed for single-phase supplies where the distributor has provided a cut-out rated up to 100 A. Engineering Recommendation P.26 is applicable to three-phase supplies and quotes somewhat higher values.

Consumer units are tested to a conditional test in BS EN 60439-3 to represent applications fed from a 100 A fused cut-out with supply fault levels up to 16 kA.

The Engineering Recommendations include information on the attenuation in fault current provided by the service cable to the consumer's installation, leading to much reduced values of short-circuit current at the service cut-out. The extent of the reduction depends on the size (csa) of cable and its length.

This reduction in short-circuit current by the service cable does not apply to installations in high load density areas of major city centres.

For other situations the distributor should be consulted or, if the source is a consumer's transformer, the manufacturer's impedance data should be applied.

The calculation of fault current at the origin of the installation, and hence of the impedance of the source, should take into account contributions, if any, from sources additional to the public or external supply. Such additional sources may take the form of local generation or of large motor loads connected electrically close to the supply point.

The lowest prospective value relates to a fault at the downstream end of the conductors protected by the device, notionally at the input to the next distribution board or at the terminals of the current-using equipment fed by that circuit.

As stressed earlier, protection of conductors in accordance with Chapter 43 may not provide protection for equipment connected to those conductors. Where a final circuit supplies a known single load it may be feasible to select a device whose characteristics will also protect the equipment, but reference to the equipment manufacturer may be necessary. Alternatively, equipment protection should be included at or in the equipment itself.

This aspect is of importance when non-fused plugs are used, such as industrial type to BS EN 60309-2. If the designer intends to incorporate such plugs he would require to establish the sizes and types of flexible cords likely to be used. There may be a case for limiting the current rating of the circuit or setting a minimum csa or maximum length of cord to be used.

Attention is drawn to Section 8 of this Guidance Note where this matter is considered further.

3.4.3 *Relocation of fault current protection*

473 Fault current protection may be located on the load side of the position required by Regulation 473-02-01 provided that the conditions given in Regulation 473-02-02 are met.

473-02-01

473-02-02

The three conditions in Regulation 473-02-02 are intended to reduce the risk of a fault occurring and, if a fault should occur, the risk of fire and danger to persons or property. The following measures are envisaged:

  (i) the use of conductor lengths which are as short as possible

 (ii) the reduction as much as is reasonably practicable of the risk of insulation breakdown, by choice of route, provision of adequate support and enclosure or barriers, together with the addition of supplementary insulation separately over each conductor

(iii) consideration of the effects of mechanical damage, including conductor displacement, either from external mechanical forces or from electromagnetic forces in the event of a fault downstream

(iv) although the prime object should be to avoid faults greater than the withstand of the conductors, locating or enclosing the conductors so that, in the event of a fault, risk of fire or danger to persons or property by the emission of flame, arcs or hot particles is prevented as far as is reasonably practicable.

Where metallic enclosures or barriers are used they must be earthed to the circuit protective conductor.

It is not satisfactory to assume that small conductors are safely expendable. If such a conductor is destroyed, the arc products will contain copious quantities of ionized gases which may facilitate the persistence and flashback of an arc to larger conductors, where arcing is almost certain to do considerable damage. Where fusing of small conductors cannot be provided the run of the conductors should be carefully designed so as to maximize the chances of containment in the event of a fault.

A small conductor, such as one supplying instrumentation from a large conductor, can be connected through a small fuse unit of adequate fault current breaking capacity mounted directly on the large conductor.

Fig 7: **interconnections between switchgear items**

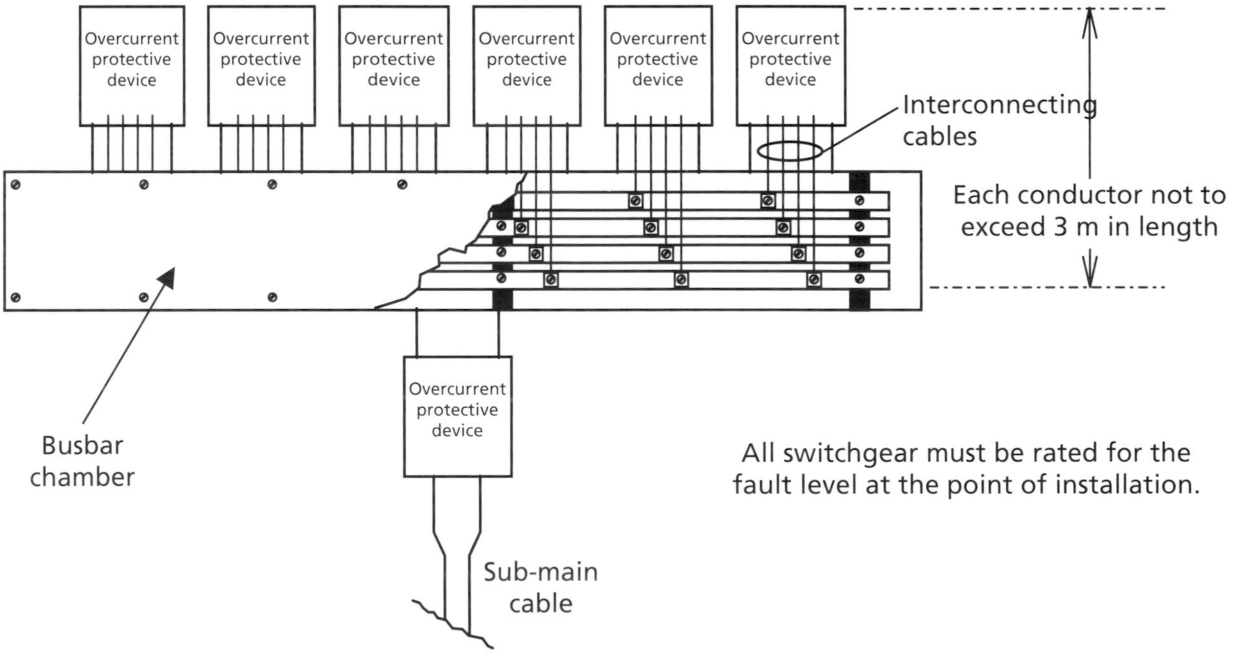

Regulation 473-02-02 is of special importance for interconnections between switchgear (Figure 7 above) and for temporary connections made to equipment where there may not be room for the installation of a suitable fault protection device at the tap-off point. In this latter situation the original integrity of the

473-02-02

50

switchgear assembly and enclosure as regards protection against entry of foreign substances, compliance with Chapter 41, Protection against Shock, and Chapter 42, Protection against Thermal Effects has to be maintained.

BS EN 60439-1 Specification for type-tested and partially type-tested assemblies, requires that the manufacturer shall specify in his documents or catalogues the conditions, if any, for the installation, operation and maintenance of the assembly and the equipment contained therein.

If necessary, the instructions for the transport, installation and operation of the assembly shall indicate the measures that are of particular importance for the proper and correct installation, commissioning and operation of the assembly.

Where necessary, the above mentioned documents shall indicate the recommended extent and frequency of maintenance.

If the circuitry is not obvious from the physical arrangement of the apparatus installed, suitable information shall be supplied, for example wiring diagrams or tables.

Regulation 473-02-03 applies where fault protection upstream of the point where a conductor cross-sectional area is reduced also provides fault protection complying with Regulation 434-03-03 for the smaller conductor. Further fault protection may then be provided at any point downstream of the conductor change (cascading), for example to provide discrimination, but not primarily to protect the conductor.

473-02-03

434-03-03

3.5 Omission of fault current protection

General

Fault current protection may be omitted altogether under certain circumstances. However, fault current protection should not be omitted, or its sensitivity reduced, without very careful consideration; see comments on Regulations 473-01 and 473-02 in paragraph 1.4 of this Guidance Note.

473

Regulation 473-02-04 allows the omission of devices for protection against fault current in the following circumstances, provided that the conductor that is not protected is installed in such a manner as to reduce to a minimum the risks of fault current and fire or danger to persons.

- For a conductor connecting a generator, transformer, rectifier or battery with its control panel.

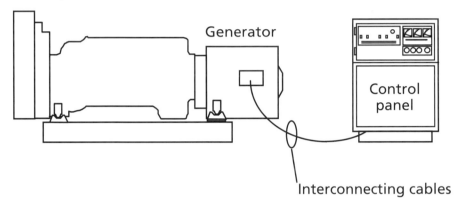

Generator

Control panel

Interconnecting cables

- In a measuring circuit where disconnection could cause danger, e.g. secondary circuits of current transformers.

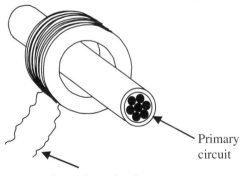

Primary circuit

Secondary circuit

The secondary circuit of a current transformer must be kept closed when current is flowing in the primary winding. If not, the peak value of emf induced in the secondary circuit will be many times the nominal value, and may be sufficient to cause breakdown of the insulation and danger.

- Where unexpected opening of a circuit causes greater danger than the fault current condition e.g. lifting magnets.

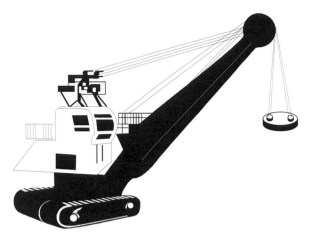

Electro-magnets are used in, e.g. scrap yards to lift and carry loads. If such a magnet is de-energised while in operation this could cause damage or injury.

- At the origin of an installation where the distributor provides a fault current device and agrees that the device affords protection to the parts of the installation between the origin and the main distribution point of the installation where the next step for fault protection is provided.

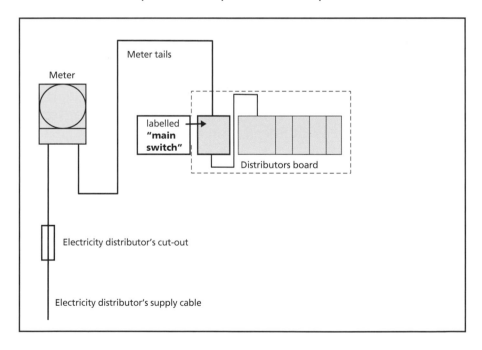

Item (iv) of Regulation 473-02-04 must be considered together with item (iv) of Regulation 473-01-04 which also deals with the protection of cables between the outgoing meter terminals and the first

473-02-04
473-01-04

point of overcurrent protection in an installation. In general, the distributor is able to make a satisfactory assessment of the maximum demand so that overload need not be a problem (see Regulation 473-01-04(ii)). The important issue will be fault current protection. Sizing and the method of installation of these cables must be agreed with the distributor. If Regulations 473-01-04(iv) and 473-02-04(iv) are not met then overcurrent protection, complying with Regulation 473-02-02, should be installed close to the origin.

473-01-04
473-02-04
473-02-02

3.6 **The use of one device for both overload and fault current protection 434**

Regulation 434-03-02 deals with a very common situation, where the same fuse or circuit-breaker provides both types of protection. The assumption relies on the inherent characteristics of the devices used, whereby correct overload protection, achieved by applying the conditions of Regulations 433-02-01 and 433-02-02, will involve devices having overcurrent characteristics which will ensure compliance with the fault current Regulation 434-03-02.

434-03-02

433-02-01
433-02-02

It may be deduced from the latter part of the regulation that non-current limiting devices may not provide the required characteristic. Doubt could arise when a device not described in Regulation 433-02-02 is used; a check must then be made, using the characteristic of the device chosen, for compliance with Regulation 434-03-03.

433-02-02

434-03-03

Fault current limitation refers to the characteristic of a device which is so constructed that interruption takes place in less than one half cycle of current. Fuses to BS 88 provide this characteristic provided that they are suitably selected with regard to the value of prospective current.

Circuit-breakers can be of either type, the current limiting type sometimes being referred to as 'fast acting'. A non-current-limiting type may be referred to as a 'zero point', 'current zero' or 'half cycle' device.

Since there is no special marking to identify these types, it is necessary to refer to the manufacturer or trade literature.

3.7   Harmonics

Regulations 473-03-04 to 07 recognize the effect of triple harmonic currents in the neutral conductor and the need to take account of this. Regulations 473-03-04 and 05 are concerned with TN or TT systems. These regulations require overcurrent detection in the neutral conductor in a polyphase circuit, where the harmonic content of the phase currents is such that the current in the neutral conductor is reasonably expected to exceed that in the phase conductors. This detection shall cause disconnection of the phase conductors but not necessarily of the neutral conductor.

473-03-04
473-03-05
473-03-06
473-03-07

The reason behind these regulations is that certain equipment such as the switched mode power supplies of computers and discharge lighting produce third harmonics. This third harmonic content produces harmonic distortion. For single-phase circuits this is not a problem from an overcurrent point of view when sizing the neutral conductor, as the neutral current always equals the phase current in both magnitude and harmonic content and no special measures need to be taken. The problem occurs in three-phase circuits. For example, 50Hz three-phase load currents, if balanced, cancel out in the neutral because of the 120 degree displacement of each phase (see figure below). Even if the load is not balanced the neutral conductor will only carry the out-of-balance current, which is less than the phase current. However, the third, and other triple harmonics combine in the neutral to give a neutral current that has a magnitude equal to the sum of the third harmonic content of each phase. The heating effect of this neutral current could raise the temperature of the cable above its rated value and damage the cable.

Therefore, consider a sub-main cable supplying a three-phase installation where there is likely to be a significant harmonic content; this is where computer loads or discharge lighting loads (or other loads which produce third harmonics) represent a significant proportion of the total load. Where the harmonic content of the phase currents is such that the current in the neutral conductor is reasonably expected to exceed that in the phase conductors, the Regulations require overcurrent detection in the

524-02-02

neutral and allowances must be made in the sizing of cables (Regulation 524-02).

More information on harmonics to enable cable sizes to be obtained is given in the Commentary on IEE Wiring Regulations 16th Edition BS 7671 : 2001.

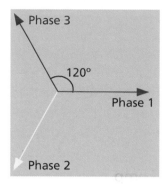

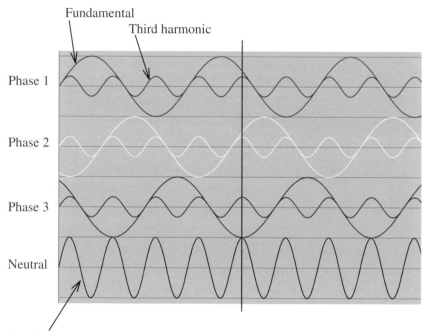

The third harmonics combine in the neutral to give a neutral current that has a magnitude equal to the sum of the third harmonic content of each phase

Section 4 — Determination of Fault Current

General 434

Regulation 434-02-01 states: The prospective fault current, under both short-circuit and earth fault conditions, at every relevant point of the complete installation shall be determined. This shall be done either by calculation, ascertained by enquiry or by measurement.

434-02-01

Unless fault current is measured directly, its determination relies on measurement or calculation of relevant values of fault loop impedances.

4.1 Determination of fault current by enquiry

Engineering Recommendation P26 published by the Electricity Association deals with the estimation of the maximum prospective short-circuit current for three-phase 400 V supplies.

Engineering Recommendation P25/1, also published by the Electricity Association, deals with the short-circuit characteristics of public electricity distributors low voltage distribution networks and the co-ordination of overcurrent protective devices on 230 V single-phase supplies up to 100 A.

For 230 V single-phase supplies up to 100 A, the electricity distribution company will provide the consumer with an estimate of the maximum prospective short-circuit current at the distribution company cut-out, which will be based on Engineering Recommendation P25/1 and on the declared level of 16 kA at the point of connection of the service line to the LV distribution cable. The fault level will only be this high if the installation is close to the distribution transformer. However, because changes may be made to the distribution network by the electricity distributor over the life of an installation, the designer of the consumer's installation must install equipment suitable for the highest fault level likely.

Since the fault level of 16 kA is at the point of connection of the service line to the LV distribution cable, a reduction of this fault level is allowed for the service line on the customer's premises. This is because the service line on the customer's premises will not be changed without the customer's knowledge. There are some inner city locations, particularly in London, where the maximum prospective short-circuit current exceeds 16 kA, and the electricity distributor would take account of this when advising the customer of the fault level.

For three-phase 400 V supplies, the electricity distribution company will provide the consumer with an estimate of the maximum prospective short-circuit current at the electricity company cut-out. This will be based on Engineering Recommendation P26, and will also be based either on the declared fault level of 18 kA at the point of connection of the service line to the LV distribution cable or on the declared fault level of 25 kA at the point of connection to the electricity distributor's substation.

Since the fault level 18 kA is at the point of connection of the service line to the LV distribution cable, again, a reduction of this fault level is allowed for the service line on the customer's premises.

In the event that the service cable is supplied from a distribution cable in the footpath on the far side of the road the attenuation in fault level can only be applied from the footpath nearest to the property. This is because the electricity distributor may, at some time in the future, install a distribution cable in the footpath nearest to the property from which to supply the service cable.

More information on the determination of fault current and detailed tables on the estimation of maximum prospective short-circuit current is given in the Commentary on IEE Wiring Regulations 16th Edition BS 7671 : 2001.

Determination of fault level at the origin by enquiry.

Figure 8 shows a number of consumers supplied via an electricity company's service cable, and the reduction in fault level at the distributors cut-outs due to the length of service cable. Shown are typical values, refer to the electricity distributor for actual values of fault level on a particular installation.

Fig 8: Reduction in fault level at the suppliers cut-outs

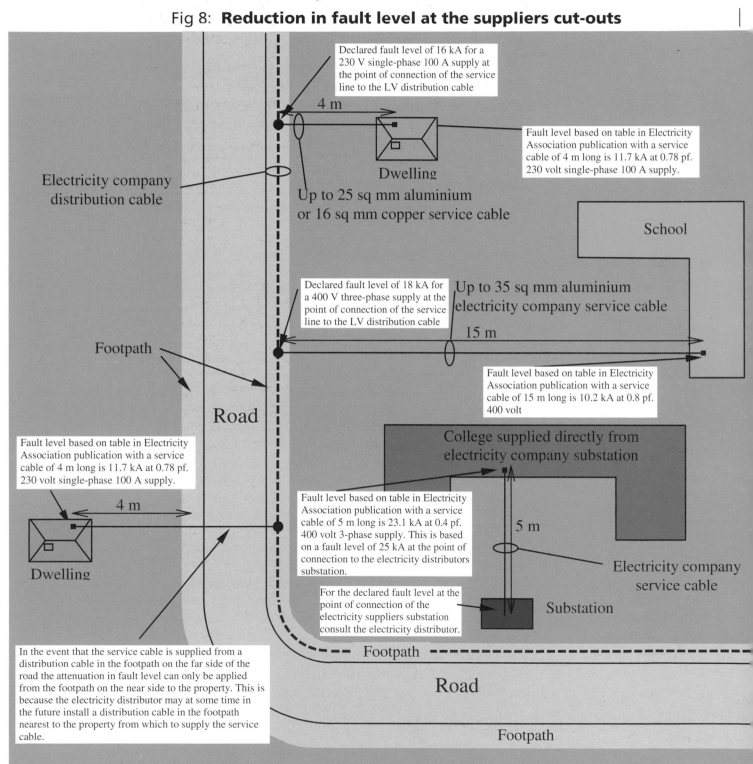

Declared fault level of 16 kA for a 230 V single-phase 100 A supply at the point of connection of the service line to the LV distribution cable

Fault level based on table in Electricity Association publication with a service cable of 4 m long is 11.7 kA at 0.78 pf. 230 volt single-phase 100 A supply.

Electricity company distribution cable

Up to 25 sq mm aluminium or 16 sq mm copper service cable

School

Declared fault level of 18 kA for a 400 V three-phase supply at the point of connection of the service line to the LV distribution cable

Up to 35 sq mm aluminium electricity company service cable

15 m

Footpath

Fault level based on table in Electricity Association publication with a service cable of 15 m long is 10.2 kA at 0.8 pf. 400 volt

Road

Fault level based on table in Electricity Association publication with a service cable of 4 m long is 11.7 kA at 0.78 pf. 230 volt single-phase 100 A supply.

4 m

College supplied directly from electricity company substation

Fault level based on table in Electricity Association publication with a service cable of 5 m long is 23.1 kA at 0.4 pf. 400 volt 3-phase supply. This is based on a fault level of 25 kA at the point of connection to the electricity distributors substation.

5 m

Dwelling

Electricity company service cable

For the declared fault level at the point of connection of the electricity suppliers substation consult the electricity distributor.

Substation

In the event that the service cable is supplied from a distribution cable in the footpath on the far side of the road the attenuation in fault level can only be applied from the footpath on the near side to the property. This is because the electricity distributor may at some time in the future install a distribution cable in the footpath nearest to the property from which to supply the service cable.

Footpath

Road

Footpath

| 4.2 | Measurement of fault current | Measurement of prospective fault current or circuit impedance requires special equipment. For circuits rated up to about 100 A, portable equipment is available and is generally used for the measurement of earth fault loop impedance. It operates by imposing a heavy load, usually greater than the rating of the circuit, for a very short duration, and measuring the consequent drop in voltage. From this data the upstream impedance and hence prospective fault current is derived. |

For higher rated circuits the currents required to effect such a measurement accurately are proportionally higher and the equipment becomes cumbersome and expensive.

To achieve a satisfactory accuracy, test currents have to be high and to avoid damage or unnecessary operation of protection devices the duration must be kept as short as possible, notionally only a few half cycles.

The measurement involves some degree of danger, increasingly so at the higher currents. It should be undertaken only by skilled personnel with adequate equipment, operated according to the manufacturer's instructions, and taking appropriate safety measures.

For a larger installation an acceptable alternative is to disconnect the circuit from the supply and to measure its impedance with reduced voltage high current test equipment. Where circuit impedance is very low this may prove to be the best, or only, way to achieve a satisfactory accuracy with safety.

With an external supply its impedance has to be ascertained from the distributor or assessed from published information and added on to the values measured within the installation.

Measurement is a useful method when there is doubt as to the status of a calculated value, where the impedance of an existing circuit is required and the route is inaccessible or, for low current final circuits, as a convenient means of regular verification of earth loop impedance. Unless a complicated method of measurement is used the resistive and reactive components of the impedance are not identified so that arithmetical addition of further impedances can only be approximate.

Accuracy of measurement, as opposed to accuracy of instrumentation, may not be as good as is sometimes thought, particularly for very extensive installations. Results may be affected by the presence of ferromagnetic materials, the effect of which can be sensitive to the value of the measuring current, near to or around conductors. This is particularly so for single-core cable circuits.

With earth fault loop measurements there is the possibility of ferromagnetic materials forming part of the loop, together with the uncertainty that subsequent operations by other trades or changes to a building structure may modify the effective path of fault currents.

Attention is drawn to the fallacy of reading the last one or two digits on a three or four digit display and omitting to overlay the result with the inherent instrument errors. The manufacturer's instructions should always be consulted before taking a reading, not only when a reading is felt to be unusual.

Where measurement indicates that the expected value of fault current is low for the type of supply concerned, consideration should be given to the possibility that the supply may be reinforced at some future date. Protective devices should be selected which will not become inadequate should reinforcement occur.

Measurement is usually done when the circuit is off-load, and the value obtained relates to conductors at ambient temperature.

4.3 Calculation of fault current 434

Derivation of circuit impedance and fault current values by calculation is the usual and sometimes the only way of providing design data. Calculation of fault current is possible only if all the impedances are known or can be estimated.

Provided that route lengths are accurate, better results are likely to be achieved with low current circuits where reactance can be ignored. Likely sources of error are connections and busbars for higher current installations, where an accurate calculation of reactance may be difficult or impracticable.

It should be borne in mind that where measurements have been made on installed circuits, values of prospective current have been reported to be lower than calculated values based on route lengths. The discrepancies have been attributed to unascertainable additional resistances and, particularly, reactances in equipment such as distribution boards and busbars. (See J Rickwood 'Fault Currents in Final Sub-Circuits' Paper at IEE/ERA Conference on Distribution, Edinburgh, 20-22 October 1970.)

Errors where actual currents are lower than those estimated may be on the safe side as far as selection of breaking capacities and withstand of electromagnetic forces are concerned, but the repercussions for thermal damage limitation, Regulations 434-01-01 and 434-03-03, are different. A lower fault current will correspond to a longer fault clearance time by the protective device and, in most cases, a greater energy let-through.

434-01-01
434-03-03

Fortunately, the situation is largely redressed by the fact that the actual operating currents of protective devices are generally somewhat lower than the maximum requirements of the relevant standards and of published data.

4.3.1

Highest value of fault current

434

The lowest impedance of a fault current path may be obtained by assuming that the fault occurs when the conductors are cold. In view of the fact that the highest value of fault current is usually required for the selection of a device with adequate breaking capacity, rather than to check on the fault current capacity of conductors, it is sufficient and on the safe side to neglect the increase in conductor resistance during a fault. Unless the conductors are located in a low ambient temperature, such as within a refrigerated area, resistance values tabulated for 20 °C may be used.

4.3.2   *Lowest value of fault current*

434
413

Contrary to expectation, the lowest value of fault
current is used when considering thermal stresses on
conductors as a result of a fault, Regulation 434-03-03.   434-03-03
This comes about because the characteristics of most
of the circuit protective devices considered in BS 7671
are such that the fault energy let-through (I^2 t) is
greater the lower the fault current. It follows that
faults are assumed to occur when a circuit is at its
maximum permissible working temperature and
conductor resistances are high.

The lowest value of fault current is also used when
confirming compliance with the maximum
disconnection times required for protection against
indirect contact (shock), and the longest likely
duration of the fault is to be checked against the   413-02-09
requirements of Regulations 413-02-09, 413-02-12 or   413-02-12
413-02-13.   413-02-13

Section 5 — Equations for the Calculation of Short-Circuit Current

5.1 General equation for fault current

The general equation for fault current is:

$$I_f = \frac{V}{\sqrt{(R^2 + X^2)}} \qquad (1)$$

where the symbols are:

I_f fault current, amperes,

V   voltage between conductors at supply point, either U_0 or U, volts,

R   resistance of the fault loop from source to point of fault, ohms,

X   reactance of the fault loop from source to point of fault, ohms.

The resistance and reactance are made up of a number of components, whose values depend on the type of fault and the supply system, see Fig 9.

Generally:

$$R = R_s + R_1 + R_n$$

$$X = X_s + X_1 + X_n$$

where the suffixes indicate:

s   supply or source quantity. Where power is obtained from a distributor these are external to the installation. It must be remembered that public electricity supply systems change to reflect growth or decline in electricity consumption for the local network and to meet day to day operational requirements. In general, a public electricity distributor cannot give a single unalterable value for the supply impedance.

1   line conductor quantity; that is, the total resistance of the line conductor(s) carrying the same fault current within the installation.

n   neutral conductor quantity, as for 1.

In most cases there is a need to establish the minimum and maximum values and these can be obtained by reference to Electricity Association publications Engineering Recommendations P23/1, P25/1 and P26. For a user owned source, a single value will generally be sufficient and this can be obtained from the manufacturer of the source equipment.

Fig 9: **General arrangement of impedances**

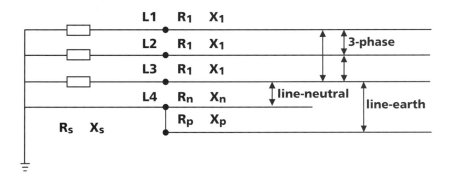

The terms R, X and Z are referred to as though they relate to a single length of conductor. It is, of course, clear that in practice they will relate to a number of conductors of different sizes in series carrying the same fault current and contributing towards the total value of R, X or Z. The value of a term in the equations given is the sum of the values for each portion of the total route from the supply point to the fault point.

This convention is of importance for final circuits fed from distribution boards, because the upstream distribution circuits may, dependent on the lengths involved, contribute significant values of impedance.

5.2   **Single-phase, line to neutral fault**

The single-phase short-circuit current for a line to neutral fault is obtained from the following equations.

5.2.1

Circuits up to about 100 A

For single-phase supplies up to about 100 A,

   (i) the separate values of R_s and X_s are not always available and an 'impedance' Z_s is substituted for R_s with $X_s = 0$.

Depending on the length of the service cable the minimum value of Z_s will be in the range 0.015 to 0.15 ohm and the maximum value generally does not exceed 0.35 ohm, except in remote rural installations with low maximum demands, where Z_s could reach 0.5 ohm or more.

GN5

   (ii) $(X_1 + X_n)$ can be neglected, so that

$$I_f = \frac{U_o}{Z_s + R_1 + R_n} \qquad (2)$$

where:

U_o is the line to neutral supply voltage

R_1 and R_n refer to the installation fault loop.

5.2.2

Circuits for more than 100 A

For circuits using conductor sizes of 35 mm^2 or larger $(X_1 + X_n)$, R_s and X_s should be included in the calculation and the single-phase line neutral short-circuit current is obtained by the use of equation (3):

$$I_f = \frac{U_o}{\sqrt{(R_s + R_1 + R_n)^2 + (X_s + X_1 + X_n)^2}} \qquad (3)$$

where:

R_s and X_s refer to a line-neutral supply loop

$(R_1 + R_n)$ and $(X_1 + X_n)$ refer to the installation line neutral loop.

5.2.3   *Sources for values of resistance and reactance*

R_1 and R_n can be calculated using conductor resistances from BS 6360 : 1991. Derivation of loop reactance $(X_1 + X_n)$ can be by any standard method, suitable equations are provided in Appendix 1 of this Guidance Note.

Alternatively, combined values of $(R_1 + R_n)$, and where applicable of $(X_1 + X_n)$, can be derived from the Tables for single-phase voltage drop for the appropriate two-core cables, or two single-core cables, in Appendix 4 of BS 7671.   Appx 4

Note that the values in the voltage drop tables are in effect expressed in impedance units of milliohms per metre at the conductor operating temperature e.g. 70 °C for pvc. For example, for use in the above equations:

$$(R_1 + R_n) = \frac{\text{Tabulated } (r) \times \text{length}}{1000} \quad \text{ohm.}$$

5.3 Conductor temperature and resistance

Resistance values taken from sources such as BS 6360   BS 6360
generally note a conductor temperature of 20 °C whereas those from Appendix 4 of BS 7671 apply to the permissible maximum operating temperature of the particular type of cable. These resistance values may need adjustment to other temperatures, depending on the assumptions to be made for the particular fault current calculation.

There may be a degree of compromise necessary in deciding on the temperature at which the resistance of each conductor in the fault loop should be calculated. (It is to be noted that circuit reactance is not temperature dependent and is not involved in these considerations.)

The following guidance on temperatures and resistances to be used for the calculation of fault current applies to the 'smallest' conductor(s) in a fault loop. Such a description includes a reduced section neutral, a protective conductor or the live conductors of the final circuit and assumes that all other conductors in the loop have equal or larger electrical cross-sections. It is accurate enough for practical purposes, and is on the safe side, if all other conductors (i.e. upstream circuits) are assumed to have resistances corresponding to their working temperatures.

5.3.1
413

Shock protection   GN5

Section 413 of the Regulations dealing with protection against indirect contact requires that the loop impedance of circuits be limited so that disconnection occurs within 0.4 or 5 s. The equation of Regulation 413-02-08 is: 413-02-08

$$Z_s \leq \frac{U_o}{I_a}$$

where:

Z_s is the earth fault loop impedance

I_a is the current causing the automatic operation of the disconnecting protective device within the time stated in Table 41A as a function of the nominal voltage U_o or, under the conditions stated in Regulations 413-02-12 and 413-02-13, within a time not exceeding 5 s

U_o is the nominal voltage to earth.

The requirement is to ensure that Z_s is of a sufficiently low value that disconnection will occur within the required time.

Regulation 413-02-05 and the notes to Tables 41B1,   413-02-05
B2, 41C, D, etc remind the user of the requirement that account must be taken of conductor temperature. The notes below the tables advise that the loop impedances should not be exceeded when the conductors are at their normal operating temperature. This means that the impedances given in the tables cannot be used for testing at cold (ambient) temperatures, but must be reduced. For example, if testing 70 °C pvc cables, the loop impedances in Tables 41B1, B2, 41C and D of the Regulations must be divided by 1.2 (see column 3 of Table 3 here) to obtain test loop impedances at 20 °C ambient.

TABLE 3
Resistance coefficients

Conductor operating temperature °C	Limiting final temperature °C	Multiplying coefficient		
		20 °C to operating temperature	20 °C to average of operating temperature and final temperature	Conductor working temperature to average of working temperature and final temperature
1	2	3	4	5
60	200	1.16	1.44	1.28
70	160	1.20	1.38	1.18
70	140	1.20	1.34	1.14
85	220	1.26	1.53	1.27
90	160	1.28	1.42	1.14
90	140	1.28	1.38	1.10
90	250	1.28	1.60	1.32

Quite clearly, reductions in supply voltage will increase disconnection times and the designer may wish to make appropriate allowances for this. Again, this is particularly important if specific device characteristics are being used, and not the worst case ones of BS 7671.

5.3.2

434

Protection against short-circuit

Regulation 434-03-03 requires that where a protective device is provided for fault current protection only (and not overload protection), it shall be ensured that the disconnection time t is such that the live conductor temperature shall not exceed those temperatures given in Table 43A. The formula quoted is

434-03-03

$$t = \frac{k^2 S^2}{I^2}$$

| 5.4 | **Single-phase circuits** | *Single-phase circuits fed from three-phase board* |

Where a single-phase circuit is supplied from a three-phase and neutral board, the portions of $(R_1 + R_n)$ and $(X_1 + X_n)$ upstream of the board, if derived from Appendix 4 of BS 7671, should be the values for single-phase circuits, i.e. for two-core cables (or two single-core cables if the upstream cables are single-core), having the same conductor size as the cable(s) supplying the board. They should not be taken from the values for three or four-core cables or circuits.

Appx 4

If the upstream cable has a reduced neutral, the mean of the tabulated values of (r) and (x) for two-core cables of each conductor size should be used.

5.4.1   *Effect of single-core cable spacing*

The reactance values in Appendix 4 of BS 7671 assume that single-core cables are installed with an axial cable spacing of either one or two cable diameters (i.e. touching or a clearance of D_e). A larger cable spacing will increase the reactance. Fig 10 gives the increase in loop reactance to be added to the term $(X_1 + X_n)$ for larger values of spacing of unarmoured single-core cables. For armoured cables reference should be made to Appendix 1 of this Guidance Note.

Appx 4

Fig 10: **Additional reactance**

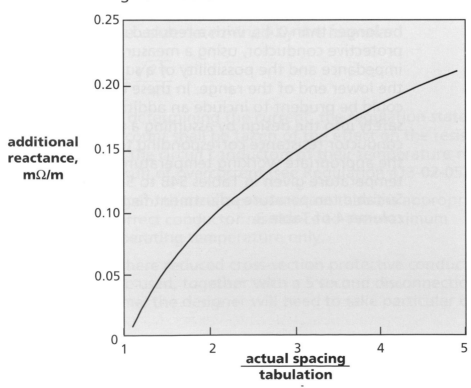

5.5 **Line to line short-circuit**

A short-circuit may occur between two lines of a three-phase circuit, in which case the current is usually less than that of a three-phase fault. It may also occur with a single-phase line to line circuit where the neutral is not distributed.

5.5.1

Circuits up to 100 A

For most practical purposes for circuits up to about 100 A, reactances X_s and X_1 can be neglected and R_s may be replaced by Z_s, where Z_s is the line to line supply 'impedance'. R_n and X_n are not involved.

The short-circuit current is then:

$$I_f = \frac{U}{Z_s + 2R_1} \tag{4}$$

where the symbols are:

U   is the line to line voltage

R_1 is the resistance of one line.

5.5.2

Circuits for more than 100 A

For circuits with conductors of 35 mm^2 or larger, the reactances X_s and X_1 should be included and equation (5) is used.

$$I_f = \frac{U}{\sqrt{(R_s + 2R_1)^2 + (X_s + 2X_1)^2}} \tag{5}$$

where:

R_s = line to line resistance of the supply

X_s = line to line reactance of the supply.

R_1 and X_1 are for one line in the installation.

5.5.3

Sources for values of resistance and reactance

For multicore cable circuits values of R_1 and X_1 can be obtained by calculation, or from BS 7671 by dividing the tabulated values for (r) and (x) for three-core cables by √3. Values of resistance may be adjusted to the appropriate conductor temperature using a coefficient from Table 3 given earlier.

For single-core cables, reference should be made to Appendix 1 of this Guidance Note, bearing in mind that for three cables in flat formation the lowest value of fault current occurs with a fault between the outer conductors.

5.6 Three-phase short-circuit 434 533

For the purpose of selection of a short-circuit protective device, for Regulations 434-03-01, 434-03-03 and 533-03-01, the three-phase short-circuit current is required. Distribution of a neutral does not affect the calculation.

434-03-01
434-03-03
533-03-01

5.6.1

Circuits up to about 100 A

X_1 is neglected, and R_s and X_s are replaced by Z_s, the line to neutral 'impedance' of the supply.

$$I_f = \frac{U/\sqrt{3}}{Z_s + R_1} \qquad (6)$$

where:

R_1 = the resistance of one line

U = the line to line voltage.

5.6.2

Circuits for more than 100 A

For circuits with conductors of 35 mm² or larger, both X_s and X_1 should be included. The fault current is then obtained from equation (7).

$$I_f = \frac{U/\sqrt{3}}{\sqrt{(R_s + R_1)^2 + (X_s + X_1)^2}} \qquad (7)$$

where :

R_s = line-neutral resistance of the supply

X_s = line-neutral reactance of the supply

R_1 and X_1 are for one line in the installation

U = line to line voltage.

5.6.3   *Sources for values of resistance and reactance*

The resistance R_1 can be calculated by reference to BS 6360. Appendix 1 of this Guidance Note provides information on the derivation of reactance X_1.

With the exception of fault loops whose impedance is attributable mainly to large single-core cables in flat formation, values of R_1 and X_1 can also be obtained from the voltage drop Tables in Appendix 4 of BS 7671 for the appropriate three-phase installation.

Appx 4

Note that the three-phase tabulated values of (r) and (x) in Appendix 4 are, in fact, $\sqrt{3}\,R_1$ and $\sqrt{3}\,X_1$, so that, in addition to the correction of (r) for temperature, both tabulated values have to be divided by $\sqrt{3}$.

Corrections to R_1 for temperature (i.e. for the highest and lowest prospective currents) apply as in the single-phase case and the same tables of coefficients can be used.

5.6.4   *Special considerations with single-core cables*

Where all of the cabling from supply to fault is multicore type, or single-core type installed in trefoil, the reactance X_1 will be the same for all phases and the fault currents will be balanced.

If a minor portion of the fault loop impedance is attributable to single-core cables in flat formation, an average value for X_1 is appropriate. This is obtained by deriving X_1 for a conductor spacing equal to 1.26 s, where s is the axial spacing between adjacent cables, see Appendix 1 or Fig 10 of this Guidance Note.

Where a substantial part of the impedance from supply to fault is contributed by a large single-core type of cable arranged in flat formation, the value of reactance for the centre cable should be used. Methods for computing this reactance are given in paragraphs 4.3 and 4.4 of Appendix 1 of this Guidance Note.

Section 6 — Equations for the Calculation of Earth Fault Current

6.1 General
543
434
413

Earth fault current has to be calculated for:

  (i) the selection of the size of the circuit protective conductor (cpc) for its limiting temperature rise

543-01-03

  (ii) checking the selection of the line conductor (in both single-phase and three-phase circuits), because the earth fault loop may have a lower impedance than the short-circuit loop

434-02-01

  (iii) checking the breaking capacity of the device intended to interrupt earth fault current

  (iv) use in connection with Shock Protection, Chapter 41, where the voltage along the cpc has to be within given limits.

413

6.2 TN systems (TN-C, TN-S and TN-C-S)

The particular feature of these systems is that there is a continuous metallic path from the exposed metalwork of equipment to the neutral of the source, see Fig 11 on next page. For such a path the value of overcurrent is high enough to operate protection such as a fuse or circuit-breaker.

fig 3
fig 4
fig 5

For reasons explained earlier, electricity distributors cannot give a single definitive value for Z_s, but have published guidance on the maximum values, which can be used to calculate minimum disconnection times, in Electricity Association Engineering Recommendation P23/1. Alternatively, a value of $(R_s + X_s)$ for the line-earth fault path of the supply for supplies up to about 100 A may be measured between terminals A and 0 with the installation disconnected. In this latter case due consideration must be given to the possibility that modifications in the supply network may change the impedance, so that consultation with the distributor is usually desirable. For user owned sources the information should be obtained from the equipment manufacturer.

Fig 11: **Earth fault path, TN-C-S system**

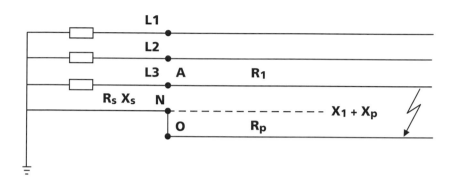

6.2.1

Circuits up to about 100 A

For circuits or supplies up to about 100 A, X_s is neglected and only R_s is used, although it may be referred to as an 'impedance'.

For such supplies typical maximum values of R_s are about 0.35 ohm if the supply is TN-C-S, or about 0.8 ohm if it is TN-S.

$$I_f = \frac{U_o}{R_s + R_1 + R_p}$$

(8)

where the symbols are:

R_1 = line resistance

R_p = protective conductor resistance

$(X_1 + X_p)$ is assumed to be zero.

This assumption depends on the two conductors being installed close to each other, which is good practice, and there being no ferromagnetic material between them. It follows that a protective conductor should be installed together with the appropriate line and neutral conductors inside any steel enclosure.

6.2.2   *Circuits for more than 100 A*

For circuits or supplies for more than 100 A the earth fault current is calculated from equation (9):

$$I_f = \frac{U_o}{\sqrt{(R_s + R_1 + R_p)^2 + (X_s + X_1 + X_p)^2}} \qquad (9)$$

where:

$X_1 + X_p$ = reactance of the line-protective conductor loop.

Generally applicable values for three-phase supplies up to 300 A are given in Electricity Association Engineering Recommendation P23/1. Alternatively, the distributor or the manufacturer of user owned sources should be consulted for values of R_s and X_s.

6.2.3   *Sources for values of resistance and reactance*

R_1 can be obtained from Appendix 9 of the On-Site Guide or Appendix E of Guidance Note 1 (at 20 °C), or from the voltage drop tables in Appendix 4 of BS 7671 (at conductor operating temperature). For the purposes of Regulation 434-03-03 the value of R_1 has to be corrected for temperature, see paragraph 5.3 on resistances for short-circuit calculations. The same coefficients can be used.

OSG Appx 9
GN1
Appx 4

434-03-03

If the voltage drop tables of BS 7671 are used the tabulated resistance values have to be divided by two in the case of values for two-core cables and by $\sqrt{3}$ for three and four-core cables, because only one live conductor is involved.

Where the protective conductor consists of a standard size conductor, the value of R_p can be obtained from the same sources as for a line conductor but using temperature adjustments appropriate to the type of protective conductor involved, see Tables 54B to 54E in BS 7671 and Table 3 of this Guidance Note.

It may sometimes be possible, where the protective conductor is reasonably close to the line conductor, to avoid computation of $(X_1 + X_p)$ by adapting the reactance for the line-neutral loop for unarmoured single-phase single-core cable in the voltage drop Tables in Appendix 4 of BS 7671. Where the protective conductor is a different size from the line conductor, the average reactance of the two conductor sizes should be used. Corrections will probably be required for a wider spacing. The procedures described for single-phase short-circuits can be followed.

Appx 4

For armoured cables and cables in conduits and trunking, where the enclosure is used as the protective conductor, special values of the loop parameters are provided below.

6.2.4   *Value of touch voltage*

413   The impedance across which a touch voltage appears is: 413

$$\sqrt{R_p^2 + X_p^2} \tag{10}$$

Where X_p has not been determined separately it can, for convenience, be assumed to be $(X_1 + X_p)/2$.

Where the size of the protective conductor is 35 mm^2 or less, X_p may be neglected.

6.3   Use of a cable enclosure as a protective conductor   Where the sheath or armour of a cable or metallic conduit or trunking is used as a protective conductor, R_p and $(X_1 + X_p)$ can be derived from the following data.

6.3.1   *Steel-wire armoured multicore cables*
543

$$(R_1 + R_p) = R_c + 1.1\, R_a \tag{11}$$

$$(X_1 + X_p) = \frac{0.3 \times L}{1000} \ \ \Omega \tag{12}$$

where:

L conductor length, m

R_c is the resistance of the line conductor

R_a is the d.c. resistance of the armour adjusted, as necessary, to the appropriate temperature. Values of armour resistance at 20 °C are given in the appropriate cable standard.

1.1 is a coefficient which represents the magnetic effect of the steel armour.

The behaviour of steel-wire armour as a protective conductor and its temperature rise with fault current is not well understood. The problem can be simplified by considering that the steel-wire behaves like a copper conductor having roughly one half of the cross-sectional area of the steel. For small cables the armour cross-section is more than adequate, while for the largest sizes its performance is nearer to that of a copper protective conductor having a cross-sectional area about 50 % of that of the phase conductors.

Selection of the appropriate resistance adjustment for temperature when calculating earth fault loop impedances (Sections 413, 434 and 543 of BS 7671) with steel-wire armoured cables follows Table 3 of this Guidance Note for copper protective conductors, but with the steel armour credited with about one half of its cross-sectional area. Armour cross-sectional areas can be obtained from BS 6346 or BS 5467.

Suitable temperature adjustment coefficients for the resistance of steel-wire armour, to be used as multipliers to 20 °C values (see BS 6345 or BS 5476), are given in Table 4 on the next page.

TABLE 4
Coefficients for steel-wire armour

Insulation material	thermoplastic		thermosetting	
	70 °C	90 °C	85 °C	90 °C
Assumed initial temperature (operating)	60 °C	80 °C	75 °C	80 °C
Final temperature	200 °C	200 °C	220 °C	200 °C
Coefficient for adjustment to operating temperature	1.18	1.27	1.25	1.27
Coefficient for adjustment to the average of the initial and final temperatures	1.50	1.54	1.57	1.54

Note: Coefficients to be applied to resistances at 20 °C.

It is important to note here that because the line conductor is contained within the armour, the latter acting as the return conductor, the reactance $(X_1 + X_p)$ is practically entirely associated with the line conductor.

The steel-wire armour on multicore cables is usually quite adequate for earth fault current duty but, in the unlikely event that this is not so, it is better to select a larger size of cable rather than to connect an auxiliary conductor in parallel with the armour. The magnetic effect of the armour will restrict the current passing through such a conductor to a small fraction of that expected on the basis of division according to d.c. resistances.

Note: Touch voltage

The impedance for the calculation of touch voltage is R_a and there is no reactive term.   413

6.3.2 *Aluminium-wire armoured single-core cable*

The armour of these cables is usually bonded together to form an earth fault return through the three armour paths in parallel. Because these paths are of low resistance, inductive effects influence the division of current between them and, except for the smaller sizes of cable, the parallel resistance is greater than $R_a/3$. Further, there is a small, but in some cases significant, reactive component to the armour impedance.

$$(R_1 + R_p) = R_c + C_r.R_a \tag{13}$$

$$(X_1 + X_p) = X_1 + C_x.R_a$$

where the symbols used are:

R_c resistance of conductor at the appropriate temperature

R_a d.c. resistance of armour of one cable at the appropriate temperature, see Table 3, Section 5 of this Guidance Note

X_1 reactance of the conductor armour loop, or the internal reactance of a single cable. A method for calculating X_1 for an isolated cable is given in paragraph 5 of Appendix 1 of this Guidance Note

C_r coefficient to be applied as a multiplier to R_a to obtain the effective resistance of three armours in parallel. The value of C_r depends on the size and arrangement of the cables

C_x coefficient to be applied as a multiplier to R_a to obtain the effective reactance of three armours in parallel. The value of C_x depends on the size and arrangement of the cables.

Values of C_r and C_x are given in Table 5 for common arrangements of single-core cables to BS 6346 and BS 5467.

TABLE 5
Values of C_r and C_x for aluminium-wire[1]
armoured single-core copper conductor cables

| Size mm²[3] | Trefoil | | 3 flat formation[2] | | | |
| | | | Touching | | Spaced[4] | |
	C_r	C_x	C_r	C_x	C_r	C_x
150	.35	.09	.35	.14	.38	.20
185	.35	.10	.35	.15	.39	.21
240	.35	.11	.36	.16	.40	.23
300	.35	.12	.36	.17	.41	.25
400	.37	.16	.39	.24	.47	.31
500	.38	.17	.39	.25	.49	.33
630	.39	.18	.40	.26	.51	.33
800	.43	.23	.45	.33	.60	.37
1000	.44	.24	.47	.34	.63	.37

Notes:

1   Values for solid aluminium conductor cables are approximately the same as those for copper conductor cables having the same diameter.

2   Earth fault assumed to be between an outer conductor and bonded armour. Values for a fault from the central conductor are slightly lower.

3   For sizes up to and including 120 mm², C_r has the conventional value of 1/3 and $C_x = 0$.

4   Axial spacing $2 \times D_e$.

The above relates to a circuit having one cable per phase with the fault at the far end of the run. Where a fault along the run is envisaged, or there is more than one cable per phase, special considerations are necessary, see Regulation 473-02-05.

473-02-05

Resistance of armour at 20 °C is available from the cable standard and can be corrected to the appropriate temperature. For calculating the highest value of earth fault current a resistance at 20 °C is suitable. When checking compliance with Regulation 543-01-03, a resistance adjustment is required (see Table 3).

543-01-03

Suitable coefficients, to be applied as multipliers to 20 °C values, are given in Table 6.

TABLE 6
Coefficients for aluminium-wire armour

Insulation material	thermoplastic		thermosetting	
	70 °C	90 °C	85 °C	90 °C
Assumed initial temperature (operating)	60 °C	80 °C	75 °C	80 °C
Final temperature	200 °C	200 °C	220 °C	200 °C
Coefficient for adjustment from 20 °C to operating temperature	1.16	1.24	1.22	1.24
Coefficient for adjustment from 20 °C to the average of the operating and final temperatures	1.44	1.48	1.52	1.48

A method for calculating X_1 is given in Appendix 1 of this Guidance Note.

Note: Touch Voltage

For the smaller sizes of cable the touch voltage is calculated using R/3, the d.c. resistance of the three armours in parallel, and the reactive term can be ignored. For larger sizes the impedance responsible for the touch voltage is

$$\sqrt{(C_r.R_a)^2 + (C_x.R_a)^2}$$

6.3.3   *Copper sheathed cables*

The approach is similar to that for aluminium-wire armoured cables, with the following modifications:

X_1 and X_p are both negligible.

For multicore cables, R_p is the sheath resistance adjusted to the appropriate temperature. For single-core cables, R_p must take account of the number of sheaths bonded together.

Resistances for conductor (R_1) and sheath (R_p) at 20 °C can be obtained from data given in BS 6207, or from the manufacturer, and adjusted to the appropriate temperatures.

To adjust resistances at 20 °C to the highest average temperature permitted during an earth fault, the coefficients in Table 7 may be used.

TABLE 7
Coefficients for copper sheaths

	70 °C	105 °C
Assumed initial temperature	70 °C	105 °C
Final temperature	200 °C	200 °C
Coefficient for adjustment from 20 °C to operating temperature	1.20	1.33
Coefficient for adjustment from 20 °C to the average of the operating and final temperatures	1.45	1.52
Value of k (Regulation 543-01-03)	1.35	1.12

543-01-03

Note: Touch Voltage

The impedance for the calculation of touch voltage is R_s, the sheath d.c. resistance, and there is no reactive term.

413

6.3.4   *Auxiliary conductors*

The use of an auxiliary conductor to supplement the fault current capacity of aluminium-wire armour or metallic sheaths calls for the placement of such a conductor as close as possible to the cable. Even when such a conductor is touching the cable its current is less than that indicated by the respective d.c. resistances. When the separation is of the order of

100 mm or more the reactive effect will reduce the effectiveness of the auxiliary conductor considerably. (The interposition of ferromagnetic material would decrease the effectiveness dramatically.)

When cables are in flat formation, the use of two auxiliary conductors placed between the cables is a way of achieving a small separation from all lines.

6.3.5 *Cables in steel conduit*

The magnetic effect of steel conduit is quite significant and affects the loop resistance and reactance. Because the effect is non-linear with current, two ranges of fault current have to be recognized in order to avoid making the design procedure unreasonably complicated. Accuracy for values of effective resistance and reactance is poor because of the variability of the magnetic properties of steel.

The effective (line + protective conductor) resistance is:

$$(R_1 + R_p) = R_c + F_r.R_{dc} \tag{14}$$

where the symbols used are:

R_c resistance of the line conductor

R_{dc} d.c. resistance of the conduit at working temperature (assumed to be about 50 °C, so that the adjustment to be applied to values at 20 °C is to multiply by 1.12)

F_r is a factor to take account of the magnetic effect of the steel.

The effective reactance of the line-protective conductor loop is given by:

$$(X_1 + X_p) = F_x.R_{dc} \tag{15}$$

where:

F_x is a factor to take account of the reactive effect of the steel.

The unexpected feature of equation (15), where the reactance is stated to be dependent on conduit resistance, comes about because the major contributor to reactance of the conductor/conduit loop is the magnetic flux within the steel. The distribution of this flux within the wall of the conduit and its value depends on the frequency of the current and the resistivity and permeability of the steel. The factor F_x is empirical, being derived from tests at 50 Hz on samples of conduit made to BS 4568. Because of the variable magnetic properties of conduit steel the values given for F_x (and F_r) should be regarded as typical.

Values of F_r and F_x are given in Table 8 for both heavy and light gauge steel conduit.

TABLE 8
Values of F_r and F_x for steel conduit

Heavy gauge conduit

Size of conduit	Fault current			
	up to 100 A		above 100 A	
mm	F_r	F_x	F_r	F_x
16	3.0	2.0	1.3	1.3
20	2.8	1.9	1.3	1.3
25	2.4	1.7	1.1	1.1
32	2.0	1.4	0.92	0.92

Light gauge conduit

Size of conduit	Fault current			
	up to 100 A		above 100 A	
mm	F_r	F_x	F_r	F_x
16	2.3	1.6	1.3	1.3
20	2.1	1.4	1.3	1.3
25	1.9	1.3	1.1	1.1
32	1.8	1.3	1.1	1.1

Note: Touch Voltage

The apparent impedance of the conduit is, for usual values of fault current, much less than its d.c. resistance. This is due to the magnetic screening effect of steel. However, for simplicity and to be on the safe side when calculating a touch voltage, it is assumed to be equal to its resistance, R_{dc}, irrespective of the value of current. There is no reactive component.

413

6.3.6 *Cables in steel ducts and trunking*

The magnetic effect of steel trunking is rather less than that of conduit, but the method for deriving values of resistance and reactance is similar. Trunking is not suitable for use as a protective conductor for circuits carrying much more than 100 A; unless particular care is taken to ensure the continuity and current-carrying capability of joints, it is possible to use only one range of fault current.

For circuits carrying no more than 100 A it is possible to derive values for $(R_1 + R_p)$ and $(X_1 + X_p)$ by one set of formulae:

$$(R_1 + R_p) = R_c + 2.1\ R_{dc} \qquad\qquad (16)$$

where the symbols used are:

R_c resistance of line conductor

R_{dc} d.c. resistance of the trunking at 50°C.

2.1 is a coefficient to take account of the magnetic effect of the steel.

$$(X_1 + X_p) = 2\ R_{dc} \qquad\qquad (17)$$

Note: Touch Voltage

The impedance to be used to calculate touch voltage   413
is R_{dc}, with no reactive term.

No protective conductor or auxiliary protective conductor should be placed outside steel conduit or trunking containing the associated live conductors. The magnetic screening effect of the conduit or trunking is so great that any external conductor will be of little effect.

6.4   TT system

The feature of this system is that there is no continuous metallic path between the exposed metalwork of the equipment and the neutral of the source, see Fig 12. Part of the return path goes via two earth electrodes and the general mass of earth. The resistance of the electrodes is usually greater than 1 ohm, perhaps up to 200 ohms, above which it is likely to be unstable.

Guidance on the installation and testing of these electrodes is given in BS 7430: 1998 Code of practice for Earthing.

Fig 12: **Earth fault path, TT system**

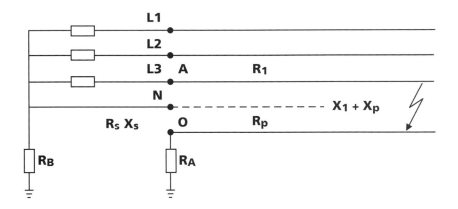

Earth fault currents are generally much lower than for TN systems. For this reason it is usually not possible to use automatic disconnection by an overcurrent device as a means of shock protection.

Calculation of earth fault current follows the same lines as those set out for TN systems, but the resistances of the earth electrodes, R_A and R_B, have to be added to the external portion of the earth fault loop, see Fig 12. Values of R_A and R_B are generally much greater than those of R_s and X_s experienced with TN systems and almost always have to be measured or estimated for each individual installation.

$$I_f = \frac{U_o}{R_B + R_A + (R_1 + R_p)} \qquad (18)$$

where the symbols used are:

R_B earth electrode resistance at source

R_A earth electrode resistance of installation.

In most installations likely to use this form of earthing with automatic overcurrent protection to comply with Regulation 413-02-19, $(R_1 + R_p)$ can usually be neglected in comparison with $(R_A + R_B)$.   413-02-19

Further, the usual value of $(R_A + R_B)$ is such that all reactances can be ignored.

Note: Touch voltage

The impedance across which a touch voltage is produced is $(R_A + R_p)$ but, as noted above, R_p can usually be neglected.   413

Section 7 — Selection of Conductor Size

7.1 General
523

Conductor sizes selected to meet the requirements for current-carrying capacity, voltage drop and, in the case of protective conductors, the requirements of Section 413, have to be checked to confirm that they will not be overheated in the event of a fault.

434-03-03
523
543-01-03

For this purpose the following regulations apply:

Short-circuit, line and neutral	434-03-03
Earth fault, line	434-03-03
Earth fault, protective conductor	543-01-03

7.2 Overload and short-circuit protection by the same device
434

Regulation 434-03-02 allows a designer to assume that if an overcurrent protective device is providing overload protection complying with Section 433, and has a rated breaking capacity not less than the prospective short-circuit current at its point of installation, it is also providing, without further proof, short-circuit protection. That is to say, there is no need to check that the circuit conductors are adequately protected thermally by verifying compliance with Regulation 434-03-03.

434-03-02

434-03-03

This assumption holds true for all fuses and energy limiting circuit-breakers. For zero point or current zero type circuit-breakers there may be a value of fault current above which the conductors are not protected. It will generally be found in practice, however, that this value is greater than the prospective short-circuit current. Advice on the type of circuit-breaker should be sought from the manufacturer (see Section 3.6 of this Guidance Note).

It is necessary to comply with Regulation 434-03-03 when short-circuit protection is provided by a separate device and to check compliance where the characteristic of the proposed device for combined overload and fault current protection is in doubt.

434-03-03

7.3 Earth fault current 543

Although the size of a phase conductor may comply with Regulation 434-03-03 under short-circuit conditions, where earth fault current is less than the short-circuit current the energy let-through of the device may be greater and it is necessary to check the sizes of both the phase conductor for compliance with Regulation 434-03-03 and the protective conductor for compliance with Regulation 543-01-03.

543-01-01
543-01-03

7.4 Parallel cables 434

The assumption of Regulation 434-03-02 should not be made for conductors in parallel. Regulation 473-02-05 sets out the requirements for the protection of cables in parallel: A single protective device may protect conductors in parallel against the effects of fault currents provided that the operating characteristic of that device results in its effective operation should a fault occur at the most onerous position in one parallel conductor. Account shall be taken of the sharing of the fault currents between the parallel conductors. A fault can be fed from both ends of a parallel conductor.

473-02-05

The regulation requires that where a single protective device may not be effective then one or more of the following measures shall be taken:

(i) the wiring is carried out in such a way as to reduce the risk of a fault in any parallel conductor to a minimum, for example, by protection against mechanical damage, and conductors are not placed close to combustible material, or

(ii) for two conductors in parallel a fault current protective device is provided at the supply end of each parallel conductor, or

(iii) for more than two conductors in parallel fault current protective devices are provided at the supply and load ends of each parallel conductor.

7.5 Use of $I^2 t$ characteristics 434

For short durations, say less than 0.1 s, the values derived from Appendix 3 of BS 7671 may be unsuitable because of the possible current limiting effect of the device. The manufacturer should be consulted as to the $I^2 t$ let-through for the expected value of prospective current.

434-03-03
Appx 3

It is then possible to substitute this value of $I^2 t$ in a re-arrangement of the equation of Regulation 434-03-03, and in Regulation 543-01-03

543-01-03

$$S = \frac{\sqrt{I^2 t}}{k} \quad mm^2$$

to obtain the smallest conductor size to comply with the maximum permitted temperature in Table 43A or Tables 54B to 54F of BS 7671.

Table 43A
Table 54B
Table 54F

7.6 Duration of short-circuit current 543

Note that there is no limitation on fault current duration inherent in the thermal requirements of either Regulation 434-03-03 or Regulation 543-01-03. While considerations of shock risk limit earth fault currents to no more than 5 s, there is no definitive limit for short-circuit currents.

434-03-03
543-01-03

The duration for which a short-circuit current should be allowed to persist is a matter of careful judgement, see Section 1.6 of this Guidance Note. Obviously, it is wise to restrict the duration of arcing, etc to an absolute minimum. However, there is no firm evidence on which to base a specific time limit for general use and consideration of this matter must take into account the particular circumstances.

The data on which the values of the coefficient k in Tables 43A and 54B to 54F of BS 7671 have been based were generally obtained with durations not greater than about 5 s in mind, but again, this is unlikely to be critical. It is nevertheless prudent to consult the cable manufacturer if a duration in excess of 5 s is contemplated.

Table 43A
Table 54B
Table 54C
Table 54D
Table 54E
Table 54F

7.7 Status of adiabatic equations 434 543

Both Regulation 434-03-03 and Regulation 543-01-03 state that the adiabatic equations they contain provide an approximate value for duration or conductor size. It should be noted straight away that

434-03-03
543-01-03

any error is on the safe side, but may be of economic interest.

The validity of the adiabatic equation depends on the extent to which heat is lost from the conductor during the period of the fault current. For a given conductor type and insulating material the heat loss can be expressed in terms of the ratio of fault current duration to the conductor cross-sectional area.

For circular conductors the limitation on accuracy can be expressed by the following examples.

For durations (in seconds) up to 0.1 x cross-sectional area (in mm^2) the loss is negligible. In practical terms this means that the adiabatic equation is of sufficient accuracy at durations up to 0.4 s for conductors of 4 mm^2 or larger, up to 1 s for conductors of at least 10 mm^2 and up to 5 s for conductors of 50 mm^2 or larger.

For most fault current interruption times the approximate nature of the adiabatic equation for circular conductors is of interest for the small conductor sizes only. Generally the economics for circuits of such sizes means that the more complicated non-adiabatic calculation is unlikely to be worthwhile.

Shaped conductors are in the size range where the approximation does not matter.

The situation changes considerably for concentric conductors such as metallic sheaths, armour and concentric neutrals. BS 7454 provides equations and data for making non-adiabatic calculations for all types of conductor and insulating material.

For bare conductors the adiabatic equation is satisfactory.

7.8 Alternative values of k 434

Regulation 434-03-03 states that the value of the coefficient k may be recalculated where the initial temperature is lower than that assumed for Table 43A. This situation arises when conductor sizes are increased above that required to comply with current-carrying capacities.

434-03-03

Such a recalculation will yield a higher value of k and permit slightly higher fault currents for a given size of conductor, the average conductor temperature should also be recalculated. The reduction in initial temperature has to be considerable for the recalculation to be worthwhile.

Appendix 2 of this Guidance Note provides the equations necessary to calculate k.

7.9 Cables in trenches

Refer to Table 4A1 of Appendix 4 of BS 7671 for a schedule of installation methods for cables including cables in trenches.

Table 4A1

Section 8 — Notes on Fault Current Withstand of Flexible Cords

8.1 General
431

Compliance with the requirements of Chapter 43 concerns only the fixed installation. The requirements do not apply to flexible cords of current-using equipment supplied from socket-outlets..

Because the size and length of a flexible cord and the nature of the load connected to a socket-outlet are not generally within the control of the installer of the fixed wiring, the subject of overcurrent protection for flexible cords is not considered to be a subject for which requirements could be included in the Regulations.

On the other hand, there are good reasons for an installer being aware of the impact of the overcurrent protection provided for a final circuit on the flexible cord which might be connected to it. The object of these Notes is to be of assistance where an installer is aware, or has control, of the cord to be used or, where this is not so, to provide guidelines for socket-outlet circuits.

Satisfactory protection of a flexible cord calls not only for the use of appropriate overcurrent protective devices but also for a restriction on length of cord. Both features depend on user co-operation or control.

The size of a flexible cord is generally chosen according to the power consumption of the equipment supplied, and often this implies quite small sizes. With small appliances convenience of handling has to be balanced against a minimum size limit based on somewhat empirical requirements for mechanical strength.

These Notes may add to the selection criteria by introducing minimum size limits based on fault current withstand.

In the realm of user choice, it may be possible to recommend limitations on length of flexible cord.

Fault current protection of a flexible cord does not necessarily imply protection for any of the circuits within connected equipment.

8.2 Influence of circuit design and length of flexible cord

The values, for a fault at the far end of a flexible cord, of short-circuit current and, in the case of a three-core flexible, earth fault current, depend on the respective loop impedances. Contributions to these impedances by the fixed installation are affected by the criteria used to provide shock protection by disconnection and for voltage drop. Fault current protection within the fixed installation may also play a part by influencing the choice of circuit conductor sizes.

On the subject of the influence of shock protection it has to be noted that Regulations 413-02-09 and 413-02-12 permit two circuit design criteria (disconnection times not greater than 0.4 s or limitation of the impedance of the protective conductor, 5 s maximum). These may result in widely different earth fault loop impedances.

413-02-09
413-02-12

The contribution by the flexible cord as with all conductors depends not only on its size but on its length. While the importance of flexible conductor size is easily recognized, the effect of length is less obvious. However, the insertion of typical values into the problem shows that it is possible to have a length of cord so great that it may result in the overcurrent protective device not operating for a very long time or, for all practical purposes, not operating at all. The danger posed by lengthy extension leads is obvious.

Although the criteria for shock protection play a part in the consideration of fault currents, the subject of shock protection itself is not within the scope of this Guidance Note.

| 8.3 | **Damage criteria for flexible cord** | *Damage to the flexible* |

8.3.1

The internationally recognized maximum conductor temperature limits for the insulating materials used in flexible cords are 160 °C for pvc and 220 °C for rubber during fault conditions. These limits correspond to a limited degree of damage, insufficient to affect the practical performance of the cord.

These temperatures assume that conductors and insulation are subject to mechanical stress, for most situations they should be conservative.

8.3.2

Acceptable length of flexible

Standards for most appliances include a requirement for a minimum length of flexible cord, where it is supplied with the appliance. A typical length is 2 m. Some small appliances have shorter lengths, while equipment such as appliances for floor treatment have much longer flexible cords.

In general, the longer the flexible cord, the lower the fault current and the greater the time taken for the overcurrent protection to operate. The energy let-through and the conductor temperature under these conditions are greater.

As a means of selecting suitable conductor sizes for various socket-outlet circuits it is assumed that unless a cord length of at least 2 m can be used, without the conductor short-circuit temperature exceeding the appropriate limit above, the size is too small. For many conductor sizes considerably longer lengths are permissible.

It can sometimes be shown that very long lengths, up to 100 m, are acceptable as regards conductor temperature, but fault durations could then be longer than about 2 min. In such situations a maximum length of flexible cord corresponding to, say, a fault duration of 5 s is suggested.

It follows that shock protection complying with Regulation 413-02-12 can be used only where lengths of flexible cords are absolutely minimal, because the regulation permits fault currents to persist for no longer than 5 s.

413-02-12

In what follows only final circuits with protection complying with Regulation 413-02-09 (0.4 s disconnection) will be considered.

413-02-09

8.4 Conclusions

As explained above, the short-circuit and earth fault duty imposed on flexible cords supplied from socket-outlets depends on the design parameters of the particular socket-outlet circuit in the fixed installation. The following conclusions are based on a survey of a range of circuit parameters, but are not claimed to be exhaustive and are provided to show the nature of the limitations involved. In many respects they confirm existing practices.

It is important to bear in mind that the conductor sizes quoted below are based on short-circuit or earth fault conditions and in no way override the requirement to observe the current-carrying capacity of the conductor and voltage drop limits for equipment. These latter requirements may often call for larger sizes.

The limitations given below are based on the worst case characteristics of fuses and circuit-breakers and, in most cases, assume that the fixed circuit just complies with the requirements of Regulation 413-02-09. It is to be expected that actual installations will impose rather less onerous duties on flexible cords so that the limitations will be conservative.

413-02-09

8.4.1 Ring circuits

For ring circuits protected by either a 30 or 32 A protective device, where the outlets are fused at 13 A or 3 A:

(i) protection of a 0.22 mm^2 conductor by a 13 A fuse against an earth fault is marginal, in that the earth fault loop impedance of the circuit must be lower than the maximum permitted by Regulation 413-02-09 (0.4 s duration)

413-02-09

(ii) 0.22 mm^2 cords are not protected where the circuit has been installed to comply with Regulation 413-02-12 (limitation of protective conductor impedance)

413-02-12

(iii) the formal position that 3 A fuses should be used in 13 A BS 1363 plugs is confirmed, especially as it is then unnecessary to restrict the length of a 0.22 mm^2 flexible cord to avoid thermal damage

(iv) however, lower impedances than those envisaged in (i) would almost certainly apply for line to neutral faults so that protection would be acceptable for two-core cords supplying Class II equipment, provided that the length of the cord does not exceed about 2 m

(v) because the risk of damage is marginal and may be negligible for some line to neutral faults on short lengths of 0.22 mm^2 flexible cord, the impression may have developed that the 3 A fuse is not strictly necessary. This is unfortunate and use of the 3 A fuse should be retained

(vi) except as noted in (i), a minimum size of 0.5 mm^2 should be required for 13 A outlets on 30 A or 32 A ring circuits. This applies where shock protection complies with either of the requirements of Section 413

(vii) pvc-insulated cords are satisfactory under the above conditions, but lengths should not extend beyond about 10 m

(viii) rubber insulated cords are preferable for extension sets with 13 A fuses. Where lengths greater than about 10 m are involved sizes should be increased to 0.75 mm^2 or larger

(ix) even if a conductor size is acceptable as regards thermal limitations, it may be that a long length has such a high resistance that a 13 A outlet fuse will not operate for a long time. A maximum length should therefore be observed so that the duration of the fault current is restricted to, say, 5 s.

8.4.2   *Radial circuits*

Radial final circuits, with no overcurrent protection at the socket-outlets.

Table 9 gives examples of maximum lengths of three-core flexible cord which have to be observed to avoid either thermal damage to the cord or an earth fault current duration exceeding 5 s.

For line-neutral short-circuits on two-core flexibles, the circuit resistance would be expected to be rather lower than that assumed above for earth faults and the lengths given could be increased.

TABLE 9
Maximum lengths of cord for earth faults (circuit protection complying with 413-02-09)

Insulated conductor size mm^2	Length m	
	pvc	Rubber
16 A fuse at distribution board		
0.5	8	14 (17)
0.75	22 (29)	22 (50)
1.0	29 (70)	29 (—)
20 A fuse at distribution board		
0.75	7	14
1.0	22	24 (53)
1.25	31 (43)	31 (70)
32 A fuse at distribution board		
1.0	N	4
1.25	5	13
1.5	14	25 (25)
2.5	40 (70)	40 (100)

N = not to be used.

Values in brackets apply to the thermal limitation and are given where the other value applies to a 5 s limitation for the fault current.

The 0.22 mm² size cannot be used at all.

Where it is not feasible to control the lengths of cord or extension used, consideration should be given to recommending minimum sizes.

Appendix 1 — Calculation of Reactance

1   *General*

It is important to bear in mind that reactance is essentially associated with a circuit or loop, generally formed by two conductors. For this reason the reactance of line-neutral and line-protective conductor loops have been expressed in this Guidance Note in the form $(X_1 + X_n)$ etc so as to avoid the impression that the contribution by each conductor can be applied in an equation, measured or considered to exist, as an independent quantity.

However, because the two conductors may be of different size or form, it is convenient for the purpose of computation to regard their reactances separately as one does the resistances. In the case of a three-phase fault the return conductor is formed by the other phases and does not appear directly in the computation.

2   *Line to neutral, single-phase, faults*

2.1   Line and neutral reactances for:

(i) all multicore cables

(ii) unarmoured or non-metallic sheathed single-core cables.

The reactance of each conductor:

X_1 and $X_n = 0.0628 \log_e(2\ s/a.d)$ m ohm/m

at 50 Hz.

where symbols used are:

s   axial separation of the two conductors, mm

d   conductor overall diameter, mm.

Shaped conductors are treated as circular conductors having the same cross-sectional area.

a   is a coefficient to allow for the internal reactance of the conductor according to Table 1.1.

TABLE 1.1

No of wires in conductors	Coefficient a
3	0.678
7	0.724
19	0.758
37	0.768
61	0.772
91	0.774
127	0.776
169	0.779

The reactance due to the helical laying up of cores of multicore cables can be neglected.

2.2   Resistance and reactance of line and neutral conductors for non-magnetic armoured or metal sheathed single-core cables.

(If the sheaths or armour are not bonded at both ends of the run, the impedance for non-metallic sheaths or unarmoured cables applies).

In this case both resistance and reactance are modified by the presence of circulating current in the armour or sheath. The apparent values of resistance and reactance of each of the conductors are given by:

R_1 and $R_n = R_c + F.R_a$

X_1 and $X_n = X_c - F.X_a$

where symbols used are:

R_c conductor resistance at the appropriate temperature

X_c reactance of each conductor calculated as in Item 2.1

R_a resistance of armour or sheath

X_a reactance of sheath or armour

$\quad = 0.0628 \log_e(2\,s/d_a)$ m ohm/m at 50 Hz

d_a mean diameter of sheath or armour, mm

s axial separation of cables, mm

F coefficient of coupling between armour and conductor,

$$= \frac{X_a^2}{R_a^2 + X_a^2}$$

3   *Line to line, single-phase, short-circuit*

3.1   Reactance of each line conductor for:

(i) multicore cable

(ii) non-metal sheathed or unarmoured single-core cables in trefoil

(iii) adjacent non-metal sheathed or unarmoured single-core cables in flat formation.

X_1 is given by the same equation as for line to neutral faults in Item 2.1, with s being the axial separation of adjacent line conductors.

3.2   Resistance and reactance for each line conductor for:

(i) non-magnetic armoured or metal sheathed single-core in trefoil

(ii) adjacent non-magnetic armoured or metal sheathed single-core cables in flat formation.

The increased resistance and reduced reactance of each line conductor are given by the equations for R_1 and X_1 in Item 2.2, with s being the axial separation between adjacent line conductors or cables.

3.3   For a fault between the outer conductors of three single-core cables of any type in flat formation, equally spaced.

The reactance and, where applicable, the resistance of each line are given by the equations in 2.1 and 2.2, according to the type of cable and installation, but with the values for both X_c and X_a increased by 0.04355 m ohm/m. In these equations the symbol s is unchanged as the axial separation between adjacent conductors or cables.

4   *Three-phase short-circuit*

4.1   Reactance of line conductor for:

  (i) multicore cables

  (ii) unarmoured or non-metal sheathed single-core cables in trefoil.

X_1 $= 0.0628 \log_e(2 \, s/a.d)$ m ohm/m at 50 Hz.

  s   is the axial separation between the line conductors, mm.

Other symbols and values of a are as before.

4.2   Resistance and reactance for non-magnetic armoured or metal sheathed single-core cables in trefoil.

The apparent increase in resistance R_1 and decrease in reactance X_1 of the line conductor are given by the equations in Item 2.2, with s being the axial separation between line conductors or cables.

4.3   Resistance and reactance for three unarmoured or non-metal sheathed single-core cables in flat formation, equally spaced.

The fault currents in the lines are not equal and the centre conductor has the lowest apparent reactance. For this conductor:

$R_1 = R_c$

$X_1 = X_c - X_m/3$

$H = X_a - X_m/3$

where:

R_c is the conductor resistance at the appropriate temperature

$X_c = 0.0628 \log_e(2 \, s/a.d)$ m ohm/m at 50 Hz, and

$X_m = 0.0628 \log_e(2) = 0.04355$ m ohm/m at 50 Hz

   s   is the axial separation between adjacent line conductors, mm.

4.4   Resistance and reactance for three non-magnetic armoured or metal sheathed cables in flat formation, equally spaced.

The apparent increase in resistance and decrease in reactance of the middle conductor is given by the following equations:

$R_1 = R_c + F.R_a$

$X_1 = X_c - X_m/3 - F.(X_a - X_m/3)$

where symbols used are:

R_c resistance of the conductor at the appropriate temperature

X_c reactance of the conductor as given for the unarmoured or non-metallic sheathed case in 4.3

X_m $= 0.04355$ m ohm/m at 50 Hz

  F   coefficient of coupling between armour and conductor

$$= \frac{H^2}{R_a^2 + H^2}$$

  H $= X_a - X_m/3$

X_a $= 0.0628 \log_e(2 \, s/d_a)$ m ohm/m at 50 Hz

  s   axial separation between adjacent cables, mm

d_a mean diameter of armour, mm.

Reactance of line and protective conductors:

    (i) where the protective conductor is provided by a core in a multicore cable

  (ii) by a separate conductor

(but not by the steel-wire armour of a multicore cable or a cable enclosure, see Section 6.3 of the main text of this Guidance Note).

The reactance of each conductor:

X_1 and $X_p = 0.0628 \log_e(2 \, s/a.d)$ m ohm/m at 50 Hz

where:

    s   is the axial spacing between the line and protective conductors, mm.

There may be different values of a and d for each conductor and values of a can be taken from Item 2.1.

Where the protective conductor is provided by a metallic sheath or by the non-magnetic armour of one cable (single-point bonding):

$R_1 = R_c$

$R_p = R_a$

$X_1 = 0.0628 \log_e(2 \, s/a.d)$ m ohm/m at 50 Hz.

$X_p = 0.$

Where the armour of single-core cables is bonded at both ends of the run, refer to paragraph 6.3.2 of Section 6 of this Guidance Note.

It is assumed that there is no ferromagnetic material between or near to the conductors.

The increase in impedance due to ferromagnetic material close to the conductors depends on the circumstances but may be of the order of 0.03 m ohm/m. It can be kept to a minimum by installing the protective conductor as close as possible to the line conductor (see Booth H C, Hutchings E E and Whitehead S, 'Current Ratings and Impedance of Cables in Buildings and Ships' *J.I.E.E. Vol. 83, No 502,* October 1938).

Appendix 2 — Calculation of k for Other Temperatures

Adiabatic Equation.

The general formula for k for given initial and final temperatures is:

$$k \ (As^{1/2}/mm^2) = K \left(\log_e \frac{\theta_1 + \beta}{\theta_0 + \beta} \right)^{1/2}$$

Where symbols used are:

θ_1 final temperature, °C

θ_0 initial temperature, °C

β reciprocal of the temperature coefficient of resistance for conductor material at 0 °C, °C

K   a constant for the conductor material, see Table 2.1.

TABLE 2.1

Material	β	K
Copper	234.5	226
Aluminium	228	148
Lead	230	41
Steel	202	78

Notes:

It should be borne in mind that the choice of an alternative value for initial or final temperature in the adiabatic equation should also be taken into account in the calculation of the average conductor temperature for which fault current is determined.

For further information on thermally permissible short-circuit currents and heating effects refer to BS 7454 (IEC 949).

Index

NOTES

NOTES

NOTES